Embryology of the Baboon

Embryology of the Baboon

Andrew G. Hendrickx

WITH CONTRIBUTIONS BY
Marshall L. Houston
Duane C. Kraemer
Raymond F. Gasser
Joe A. Bollert

Illustrations by Gerard T. Rote, Jr.

The University of Chicago Press

Chicago and London

This research was sponsored by the National Institutes of Health under grants number GMFR 13252 to Southwest Foundation for Research and Education, San Antonio, Texas, and RR 00169 to the National Center for Primate Biology, University of California, Davis, California.

International Standard Book Number: 0-226-32712-4

Library of Congress Catalog Card Number: 73-138518

THE UNIVERSITY OF CHICAGO PRESS, 60637
The University of Chicago Press, Ltd., London

Contents

Foreword

This book represents an endeavor to set forth in a straightforward manner the embryology and placentology of a nonhuman primate, the baboon (*Papio sp.*). At this time as our horizons of knowledge expand, man is still the most studied animal from all biomedical aspects, and embryology is no exception. Except in rare cases, the human cannot serve as an experimental animal for investigating the etiology of his own embryopathies, or for studying the relation of his normal developmental morphology to the intricacies of biochemical and physiological functions. It therefore behooves the scientist to seek out and develop animal models with demonstrated parallelisms to the human. This book is a milestone in that it represents the first monograph in primate embryology. Of the 23 Streeter stages studied (representing about 47 days of development), Stages IV through IX have never before been fully depicted, although Hertig and his associates had previously surveyed 34 human embryos for Stages I through VII. Stages beyond XXIII are in the realm of fetology and are treated succinctly but thoroughly in the Appendix as a survey of fetal growth. The characteristics used to follow the Streeter staging are the simple structures such as the eye, ear, lungs, and other derivatives of the foregut. The developing heart and brain are, for the most part, excluded from this consideration because it was felt that the heart does not provide a good staging criterion and because the brain past the development of the 5 vesicles (Stage XIV) becomes too complex for a general survey. Of course, it is well recognized here that Streeter stages are all characteristic of human embryos and that therefore an inevitable bias is established when depicting the developmental stages of another mammal, whether it be mouse or baboon. Moreover, correlations between morphological development and age are not really possible since Streeter used the ovulation age of the rhesus monkey to follow the development of and establish the staging in man. Although little comparison with human fetal development is attempted, this is unnecessary since for the most part close parallelisms to human development exist throughout mammalian species. Only in those few instances wherein acute differences exist are such variants documented and then without rationale or speculation.

Since illustrations play a critical role in the proper presentation of the interrelations among growing parts, all graphic representations are from actual wax reconstructions and slides without interpretation or artistic license.

A vast majority of the material in this book comprises new contributions by the author and his associates. The monograph itself represents a foundation upon which many complementary and tangential fields in the biomedical sciences will build for a better understanding of the life processes before birth.

LEONARD R. AXELROD

vii

Preface

In the past few years, scientists have focused attention and interest on nonhuman primates for two important reasons: to learn more about the normal biology of these species and to determine their possible usefulness as models for research on the many health problems that plague man. The purpose of this book is to emphasize specific aspects of development of a nonhuman primate, the baboon, which has not been done before, as well as to present the results of research in the related areas of reproduction and laboratory methodology. The book is designed to assist investigators in embryology, teratology, reproductive biology, and primatology.

This monograph is divided into nine chapters and an appendix, each concerned with a different aspect of development. Chapter 1, "Reproduction," contains a general description of the anatomy of the female reproductive organs, vaginal exfoliative cytology, and the menstrual cycle. Data on breeding and reproductive performance, important to a study of normal embryology, is also included.

Chapter 2, "Methods," was included to explain the recording, collecting, processing, reconstructing, and photographing techniques used in studying the embryos.

Chapters 3 through 8 all deal with stages of baboon development. A total of 23 developmental stages are described and illustrated, beginning with Stage I, the 1-day-old fertilized ovum, and ending with Stage XXIII on the 45th to 50th day of gestation when the long bones begin to ossify. To make the developmental stages as valuable as possible for comparative purposes, the stages were patterned after Streeter's "Developmental Horizons in Human Embryos" (a series of five articles in *Contributions to Embryology,* vols. 30–34, 1942–51). Consequently, each chapter is subdivided into three parts; Age and Size, External Characteristics, and Internal Characteristics. The final placement of an embryo into a stage is based on its internal characteristics; in Stages IV through VIII the trophoblast and placenta are included as staging characteristics. The average estimated fertilization age and crown-rump length for each stage are listed at the beginning of each chapter and the pertinent data for each embryo are incorporated into a table for each chapter.

The description of developmental stages I through IX (chapters 3–5) is unique since these stages have not been described in definitive form for man; Hertig and co-workers, however, have contributed greatly to our understanding of this phase of human development.

Every effort was made to determine the actual or true age (estimated fertilization age) of the embryos for a given stage by basing it on matings limited to a short duration of approximately 8 hours and considering factors such as sperm and ova viability, ovulation, and optimal mating time. Variation in the age of embryos with the same level of morphological development is also elucidated.

Chapter 9, "Placenta," deals with the morphology of the developing and term placenta and with placental circulation. It also includes a short section on the comparison and classification of the primate placenta.

Each chapter has its own list of references, selected to support statements made in the text or for their review of the subject and additional pertinent literature.

The extensive illustrating includes a large number of photographs and photomicrographs, used in support of observations described in the text. The magnifications of the photomicrographs are given in the legends. Where possible, magnifications for specific organs are uniform from stage to stage and chapter to chapter to allow comparisons as well as to maintain some degree of consistency. Drawings are used to illustrate only when photography was impossible or when, in a very few instances, suitable specimens were not available.

I am especially grateful to the National Institute of General Medical Sciences, Division of Research Resources, and to the Bureau of Health Professions, Education, and Manpower Training of the National Institutes of Health for their generous support of this five-year study. To Dr. Harold Vagtborg and Dr. Leonard R. Axelrod I owe a special debt for their encouragement in undertaking this work. I am grateful to the contributing authors for their many contributions and fine cooperation. I wish to thank the veterinary staff of the Southwest Foundation, especially Drs. Robert L. Hummer, Thomas Vice, and Franklin Kriewaldt for the part they played in managing the breeding colony and performing the surgery, often at undesirable hours. I am particularly grateful to Geraldine Sanchez, Cornelio Celaya, Guadalupe Bueno, Esther Rodriquez, and Barbara Richardson for their excellent technical assistance throughout this study. I wish to thank Patrick Click and Bill Grasser for their assistance in preparing the photographs and graphic illustrations.

My grateful thanks are extended to Betty Haas, Carol Lambert, Norma Clute, and Laurie LaBove of the Southwest Foundation secretarial staff and to Roslyn Jund and Angelina Castillo of the National Center for Primate Biology, University of California, Davis, for their assistance in completing the manuscript.

I wish to acknowledge the support of the National Center for Primate Biology, University of California, Davis, during the final preparation of the manuscript.

My thanks are due to the staff of the University of Chicago Press for their competence and their patience.

Lastly, I must pay tribute to my wife and family, because the time that I have given to writing this monograph is really their time. Their encouragement and understanding made its completion possible.

A. G. H.

Embryology of the Baboon

1

Reproduction

Andrew G. Hendrickx/Duane C. Kraemer

I. Introduction

The baboon is being used more frequently as a laboratory animal, a position it gained recently with the discovery that atherosclerosis is a naturally occurring disease in this primate. Earlier, Gillman and Gilbert (1946) of the University of Witwatersrand, Johannesburg, South Africa, used the baboon successfully for many years in their reproductive physiology and endocrinology studies. Since this introduction into the laboratory, baboons have been studied and used in many areas of science. Its relatively high position on the phylogenetic scale, large size, hardy constitution, and adaptability to life in captivity have made it the choice of many investigators.

Opinions regarding the taxonomic status of the genus *Papio* vary between extremes of recognizing five distinct species to combining them all under the single polymorphic heading of *Papio cynocephalus* Linn, 1776 (See Hill 1967). In view of these varied opinions, the authors are adopting the nomenclature proposed by Maples (1967), who has done extensive field studies on the Kenya baboon in an attempt to clarify the taxonomic relationships. He said, "it is suggested from the nature of the relationship of the populations in Kenya, the Buettner-Janusch's contention is correct that all baboons other than *Papio hamadryas* represent a single polytypic species, *Papio cynocephalus*." This then would include within the single species *P. cynocephalus* all the baboons imported from Kenya which were used at the Southwest Foundation for Research and Education, San Antonio, Texas. Previously these animals had been separated into two species, *P. anubis* (also known as *P. doguera*) and *P. cynocephalus*.

II. Female Genital Organs

The female genital organs are divided into an internal and an external group. The internal organs consist of the ovaries, uterine tubes (Fallopian tubes or oviducts), uterus, and vagina. The external organs include the vestibule of the vagina, clitoris, labia, ischial callosities, and the perineum; the last two are treated with general comments here.

A. *Internal Genitalia* (Fig. 1.1)

1. *Ovaries.* The ovaries are two almond-shaped, nodular bodies (fig. 1.1*A*), one located on each side of the uterus on the dorsal side of the broad ligament

3

within a bursa formed by the caudal portion of the mesosalpinx. They become displaced with various postures. Each ovary is attached to the uterus by a short ovarian ligament and to the uterine tube by the ovarian fimbria. The mesovarium is a fold of peritoneum that joins the ovary with the broad ligament. The suspensory ligament, a continuation of the broad ligament, courses cranially toward the kidney and contains the ovarian vessels. The ovaries measure approximately 14 mm in length, 9 mm in width, and 10 mm in thickness.

2. *Uterine tubes.* A uterine tube is located on each side of the uterus and passes from the lateral margin of the uterus where the fundus joins the body. It is directed dorsally for a short distance and then cranially and laterally, looping around the ovary. The cranial portion of the broad ligament suspends the tube and is called the mesosalpinx. The mesosalpinx is divided into two portions, one cranial and the other caudal to the tube. Each tube is about 55 mm long and

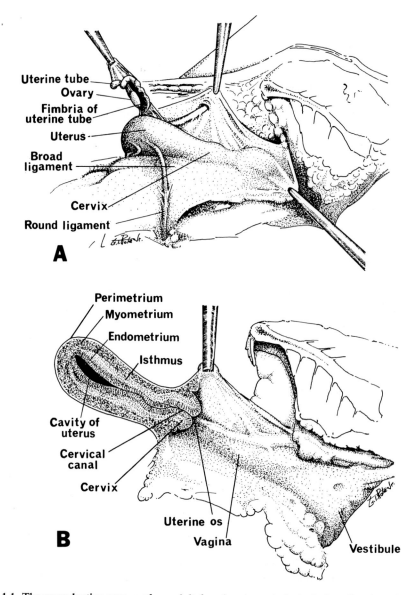

Fig. 1.1. The reproductive organs of an adult female: *A*, ventrolateral view; *B*, midsagittal view.

consists of three portions: (1) the isthmus, or constricted one-third proximal to the uterus, a portion of which courses within the visceral peritoneum of the uterus; (2) the ampulla, or intermediate dilated portion which curves over the ovary; and (3) the infundibulum, which possesses fimbriae at its ovarian end and is attached to the ovary by the ovarian fimbria. The uterine opening of the oviduct is small but the abdominal opening is considerably larger.

3. *Uterus.* The uterus is a hollow, pear-shaped, muscular organ of the simplex type (fig. 1.1*A*). It is located in the pelvic cavity between the bladder and the rectum and can be subdivided into four parts: (1) the *fundus,* which lies cranial to the entrance of the uterine tubes, (2) the *body,* which contains most of the cavity and narrows caudally into (3) the *isthmus,* and (4) the *cervix* or the portion between the vagina and the isthmus. The uterus measures about 40 mm in length from the fundus to the isthmus, 30 mm in width, and 25 mm in thickness. During pregnancy the uterus becomes enormously enlarged, attains the length of 170 mm or more, extending into the epigastric region, and measures about 115 mm in width. After parturition the uterus returns to almost its original size.

The cavity occupies only a small portion of the uterus. It is flattened dorso-ventrally into a narrow slit with the dorsal and ventral walls closely apposed. In the longitudinal plane the cavity is triangular shaped with an orifice at each angle. Cranially, the uterine tubes open into the cavity from each side between the fundus and the body. Caudally, the cavity narrows at the isthmus and is continuous with the cervical canal. The cervical canal is rather tortuous, coursing ventrally, then dorsally, and finally ventrally again (fig. 1.1*B*). It terminates at the uterine os where it communicates with the cranial portion of the vagina.

The uterine wall is subdivided into three layers: (1) the endometrium or tunica mucosa, which lines the uterine cavity, (2) the myometrium or tunica muscularis, the thickest layer, and (3) the perimetrium or tunica serosa, the thinnest layer, which is derived from the peritoneum and covers most of the uterus.

The uterus is not firmly attached or adherent to any part of the skeleton but is suspended in the pelvic cavity by ligaments. The paired, broad ligaments are thin, fibrous sheets which extend laterally from each margin of the uterus to the wall of the pelvis. The round ligaments are two flattened, fibromuscular cords attached to the cranial part of the lateral margin of the uterus, caudal and ventral to the uterotubal junction. It traverses the pelvis and penetrates the abdominal wall through the internal inguinal ring. The fundus of the uterus is inclined ventro-cranially and the uterine os is directed dorsocaudally. The position of the uterus is altered by a full bladder or a distended rectum. A full bladder tilts it dorsally; a distended rectum, ventrally.

4. *Vagina.* The vagina is a very dilatable musculo-membranous canal located ventral to the rectum and dorsal to the bladder. It extends ventrally and caudally from the uterus to the vestibule and is constricted near the vestibule and dilated in the middle. Its diameter at the vestibule is approximately 15–30 mm. The length of the vagina is about 65 mm.

The upper part of the vagina surrounds the vaginal portion of the cervix and contains recesses or fornices. Since the cervix is directed in a dorsocaudal direction, the recess dorsal to the cervix (dorsal fornix) is slightly larger than the recess ventral to the cervix (ventral fornix).

The vaginal wall is composed of an inner mucous layer surrounded by a muscular coat. A thin layer of areolar connective tissue separates them. The mucous layer lines the vaginal canal and is continuous with the uterine lining at the uterine os. Its surface is thrown into folds, or rugae, of variable sizes which extend both longitudinally and transversely.

The mucous lining of both the vagina and the uterus undergo periodic buildup followed by breakdown with sloughing of cells in response to changes in the hormone balance. These cyclic changes will be discussed under section III, *Vaginal Cytology,* and subsection III, C, *Menstrual Cycle Phases,* respectively.

B. *External Genitalia* (Fig. 1.2)

1. *Vestibule of the vagina.* The vestibule (see also fig. 1.1*B*) is a funnel-shaped, moderately shallow recess between the labia which opens externally ventral to the anus. Both the vestibule and the anus are surrounded by extensive sex skin. The vagina and the urethra open into the vestibule. The urethral orifice is ventral to the vaginal orifice and is surrounded by two papillary folds. Because of its location in the depths of the vestibule, it is concealed from view externally. However, a catheter can be inserted into the urethra without the aid of a speculum.

2. *Clitoris.* The clitoris is a rather pronounced body in most baboons and is located in the midline ventral to the vestibule. It is just ventral to a plane through the ischial callosities and is frequently masked by the clitoral lobes during maximum perineal turgescence.

3. *Labia.* The labia are poorly developed, longitudinal folds, one on each side, just lateral to the vestibule and clitoris. They are not divided into major and minor portions. During the menstrual cycle the labia undergo changes. Between the labia is a fissure, the rima pudendi, which is an irregular shallow slit with puckered edges during perineal quiescence, but during perineal turgescence it becomes a deep slit with smooth walls, extending from the anus to the clitoris.

4. *Ischial callosities.* The ischial callosities are two hairless areas of thick cornified skin that overlie the ischial tuberosities. They vary in shape from round to slightly ovoid and lie on each side of the external genitalia.

5. *Perineum.* The perineum is a region at the caudalmost part of the trunk that is bounded dorsally by the base of the tail, ventrolaterally by the ischial callosities and ventrally by the pelvic arch. The anus and external genital organs are located within this region, and the anal canal and vaginal vestibule communicate with the exterior. The perineal body is located between the vestibule and the anus. For the most part, the skin of the perineum is hairless and is very distensible but easily lacerated. It stretches to a remarkable degree during certain phases of the menstrual cycle.

The character of the perineal (sex) skin follows one of two different patterns especially evident during the turgescent phase. Most often the sex skin is smooth and unlobulated, rarely extending laterally beyond the outer border of the ischial callosities (fig. 1.2*A*). There is no swelling of the hairless buttocks. A second pattern has lobulated sex skin with the swelling extending laterally, into the hairless buttocks and occasionally even onto the flank (fig. 1.2*B*).

III. Vaginal Cytology

The cyclic changes in the baboon vagina may be studied by two methods: vaginal smear and vaginal biopsy. Variations in the thickness of the vaginal epithelium are determined best by securing vaginal biopsies. Minute pieces of the mucosa are removed by means of a suitable instrument, then sectioned and stained.

For routine determinations of vaginal changes, the smear method is as applicable to the baboon as it is to rats, guinea pigs, and menstruating women. Vaginal smears can be taken by inserting a cotton swab, previously dampened in physiological saline, into the cranial one-third of the vagina, known as the *zone of election*. The cotton swab is rotated along the lateral surface of the epithelium and then withdrawn. It is rolled onto a clean glass microscope slide, which is immediately placed into a fixative consisting of equal parts of 95% ethanol and ether. After fixation for a minimum time of 15 minutes, the smears are stained according to the method of Papanicolaou (1963).

After staining, a differential cell count is made for the purpose of estimating the effects of hormones on the vaginal epithelium. Four different cell types—the

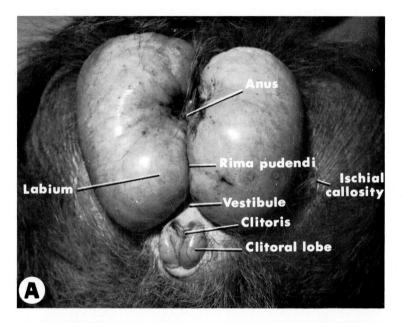

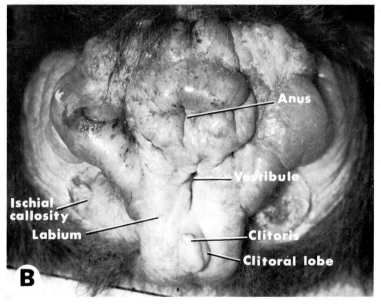

Fig. 1.2. The perineums of two normally cycling adult females on the 1st day of deturgescence. Note the variations between the two patterns. *A,* the smooth, unlobulated labia are divided medially by an even rima pudendi and are separated ventrally from the clitoral lobes by deep grooves. The swelling does not extend laterally to the ischial callosities. *B,* the labia are lobulated with indistinct boundaries on all surfaces and the swelling extends laterally beyond the ischial callosities.

superficial, intermediate, parabasal, and basal—and the occurrence of erythrocytes, leukocytes, mucus, and bacteria characterize the changes for each cycle. The following description of the cell types is based on smears taken from normally cycling proven fertile females (Hendrickx 1967).

A. *Cell Types*

1. *Superficial cells* (fig. 1.3*A–C*). The superficial cells are the largest cells seen in vaginal smears and are characterized by their pyknotic nuclei. They are derived from the stratum corneum and are usually polyhedral with clearly defined cell borders which may be irregular or indented. Their cytoplasm is delicate, light, and transparent, and is eosinophilic or cyanophilic. As the squamous epithelium matures, the nuclei become pyknotic and the staining property of the cytoplasm changes from cyanophilic to eosinophilic. This maximum epithelial maturation is indicated by the presence of large eosinophilic superficial cells with pyknotic nuclei. The pyknotic nuclei stain a deep black or violet and show no discernible structural detail. Fine or coarse granules are often present in the cytoplasm (fig. 1.3*A, B*). They stain the same color as the cytoplasm or appear brownish black. They concentrate near the nucleus and occur most frequently during the latter part of the ovulatory phase of the cycle.

Another form of the superficial cells is the keratinized or anuclear type (fig. 1.3*C*). Such cells stain pink to tannish orange and the nuclei are only faintly outlined or absent. Their origin is difficult to determine but they probably arise from the vulva and are carried to the vagina when the smear is taken or perhaps during copulation.

2. *Intermediate cells* (fig. 1.3*D, E*). Intermediate cells are derived from the stratum spinosum superficiale. They are characterized by a vesicular nucleus, which exhibits structural detail, and cyanophilic cytoplasm, although it is not unusual to see eosinophilic cells (fig. 1.3*D*). Both small and large intermediate cells are recognized. The small intermediate cells are polyhedral, usually larger and slightly flatter than parabasal cells. The nuclei are relatively large and vesicular and may be eccentric. The cytoplasm shows a bluish staining reaction. The large intermediate cells resemble superficial cells in shape and size. They differ from them only by the vesicular nucleus. The transparent cytoplasm is usually cyanophilic but may be eosinophilic.

3. *Parabasal cells* (fig. 1.3*E, F*). Parabasal cells are derived from the stratum spinosum profundum of the vaginal epithelium. They show a great variety of shapes but are usually rounded. The cytoplasm usually takes up the cyanophilic stain, but eosinophilia occurs occasionally. The cytoplasm stains less intensely than that of the basal cells and may contain vacuoles of variable sizes. The nucleus is large, round, and centrally located with a delicate but distinct structure. The eosinophilic cells mentioned above usually have a pyknotic nucleus.

4. *Basal cells.* The basal cells are the smallest cells of the vaginal epithelium and originate from the deepest layer, the stratum cylindricum. They are usually rounded or oval when observed in the smear. The cytoplasm stains a darker blue than the other cells. The nucleus is relatively large, round or oval. It is vesicular and lies in the center of the cell. These cells rarely occur in the vaginal smear of the baboon.

B. *Cell Indexes*

One hundred or two hundred cells are counted for each smear. Three indexes, the *karyopyknotic index,* the *maturation index,* and the *maturation value,* are used

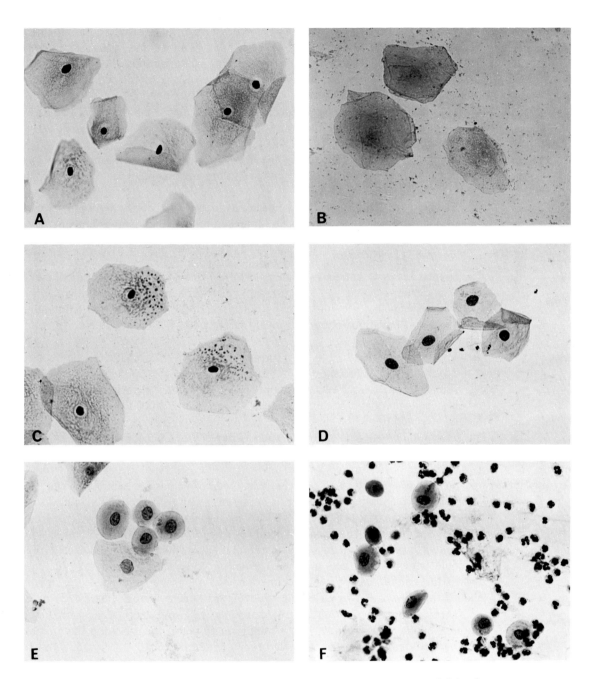

Fig. 1.3. Normal vaginal epithelial cells found in vaginal smears. (×350) *A,* superficial cells. Note the pyknotic nuclei and perinuclear vacuoles. *B,* anucleated superficial cells. *C,* superficial cells containing cytoplasmic granules. Brown tan cytoplasm also characterizes this cell. *D,* large intermediate cells. Note the vesicular nuclei and the transparent cytoplasm. A perinuclear vacuole is developing in the cell to the right. *E,* intermediate and parabasal cells. The three parabasal cells (*upper center*) and the small intermediate cell (*upper left*) are connected by intercellular bridges. Note that the nuclei of both cell types are vesicular and that the nuclei of the parabasal cells are about the same size as the intermediate cells. The sex chromatin mass is visible in the nucleus of the large intermediate cell. *F,* parabasal cells. The chromatin granules are distinct. Note the leukocytes.

to evaluate the hormonal activity. The karyopyknotic index expresses the number (percentage) of mature superficial cells in relation to mature intermediate cells regardless of staining reaction. The maturation index expresses the relation of parabasal cells to intermediate cells to superficial cells. It is expressed in percentages such as: 0/70/30, indicating 0% parabasal, 70% intermediate, and 30% superficial cells. In expressing the cytological changes as a maturation value, a numerical value is assigned to each of the following cell types: superficial cells, 1.0; intermediate cells, 0.5; parabasal cells, 0.0.

When the percentage of each of these cells present in a smear is multiplied by the corresponding numerical factor, the resulting number between 0 and 100 is the maturation value.

Erythrocytes, leukocytes, mucus, bacteria, sperm, and debris are recorded using a 6-point scale; 0 indicates a complete absence and 5 the maximum amount.

C. *Menstrual Cycle Phases*

If meticulous attention is paid to detail, 7 phases can be described for a complete ovulatory cycle: (1) menstrual, (2) postmenstrual, (3) preovulatory, (4) ovulatory, (5) postovulatory, (6) luteal, and (7) premenstrual. The range and length of the various phases, given in parentheses, is considered average for a normal 34-day cycle.

1. *Menstrual phase* (1–6 days, average 3 days). The characteristic feature of this phase is the occurrence of erythrocytes (fig. 1.4*A*). Epithelial cells, leukocytes, mucus, and cellular debris comprise the remaining picture. Well-preserved red blood cells predominate during the early stages, but in the later stages they show marked degeneration. The degenerated red blood cells may be irregularly shaped and only faintly outlined, and may stain either acidophilic or basophilic. Mucus, which stains blue, appears throughout this phase but increases during the last few days. The leukocytes, mostly of the polymorphonuclear type, are evenly scattered, unless they are trapped in the mucus, and tend to increase in number during the latter half of this phase. Cellular debris is present along with bacterial bacilli, histiocytes, and endometrial cells.

The red blood cells and mucus tend to obscure the epithelial cells, which are represented by cells from all zones of the epithelium. The intermediate cells predominate but parabasal and superficial cells are present. Superficial cells increase toward the end of the phase. The intermediate and superficial cells are slightly smaller than those that occur in the preovulatory and ovulatory phases.

Menstruation is often difficult to detect by the usual external signs of blood on the genitalia; consequently, if it is important to detect menstruation the animals are swabbed daily during the suspected time of menstruation. A few animals men-

---

Fig. 1.4. Normal cyclic changes in vaginal smears. ($\times$175) *A*, menstrual phase, 1st day of cycle. Erythrocytes predominate. A few leukocytes and small intermediate cells are present. *B*, postmenstrual phase, 4th day of cycle. Small and large intermediate cells, most of them with vesicular nuclei, are moderate in number and well dispersed. *C*, preovulatory phase, 6th day of cycle. Large intermediate and superficial cells predominate. The nuclei of the superficial cells are becoming pyknotic. *D*, ovulatory phase, 11th day of cycle. Superficial cells, containing granules, perinuclear vacuoles and pyknotic nuclei, are abundant. *E*, ovulatory phase, 17th day of cycle. Intermenstrual bleeding is indicated by the presence of erythrocytes among the superficial cells. *F*, postovulatory phase, 19th day of cycle. Superficial cells predominate but large intermediate cells are present. Clumping, folding, and curling of the cell's edges are prevalent. *G*, early luteal phase, 22d day of cycle. Leukocytes appear in moderate numbers among the folded and clumped intermediate cells. *H*, luteal phase, 26th day of cycle. Leukocytes are abundant among the large intermediate cells, small intermediate cells, and mucus. *I*, premenstrual phase, 30th day of cycle. Leukocytes, intermediate cells, and mucus are present.

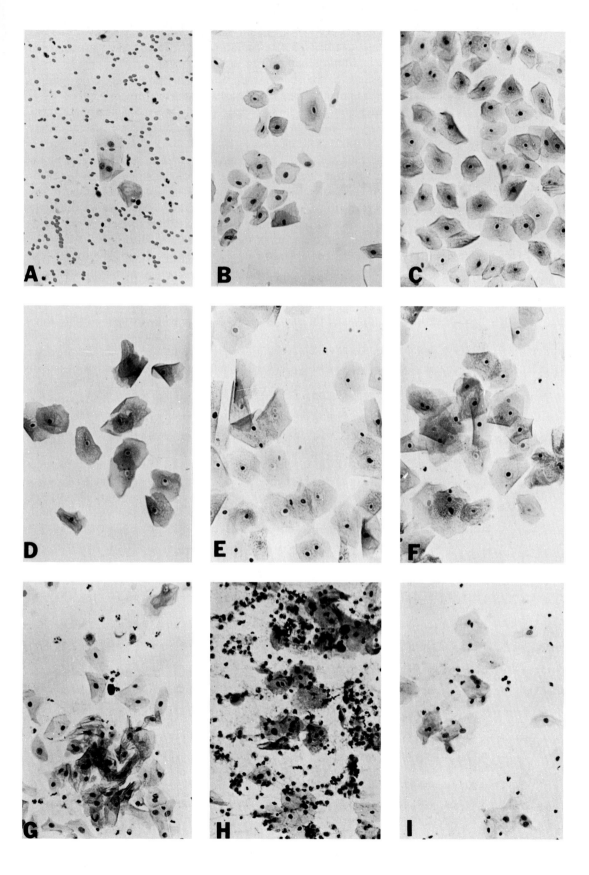

struate profusely, leaving large pools of blood and debris in the cage trays. In contrast, others menstruate so little that the only means of determining this phase is by identifying the scattered red blood cells in the smear.

The gross appearance of the stained slide during this phase is a dirty reddish color due to the large number of erythrocytes and extensive mucus.

2. *Postmenstrual phase* (2–6 days, average 3 days). During this phase there is an increase in both small and large intermediate cells and to a lesser extent superficial cells (fig. 1.4*B*). Parabasal cells occur occasionally and the leukocytes decrease in number. The cells are spread more uniformly than in the menstrual phase and the smear is relatively thin. Leukocytes occur occasionally and are widespread.

A note on the relative abundance of leukocytes is appropriate here. If the number of leukocytes tends to be high through the proliferative portion of the cycle it will remain high, with leukocytes occurring in small numbers during the ovulatory phase. On the other hand, if the smears of an individual are leukopenic during the proliferative portion of the cycle, leukocytes will disappear entirely during the ovulatory phase and may reappear in large numbers in the second half of the cycle. This fact has been previously witnessed by De Allende and co-workers (1943) in both the human and the monkey.

The gross appearance of the stained smear has a bluish cast and is quite thin.

3. *Preovulatory phase* (2–6 days, average 4 days). The abundance of superficial cells and a sharp rise in the quantity of leukocytes and mucus over the previous phase characterize the preovulatory phase. The sudden rise in the number of leukocytes and the increase in mucus again gives the smear a heavy bluish cast, contrasting with the preceding and succeeding phases and thus identifiable in the laboratory. Both the superficial and intermediate cells are large, and in most cases are relatively isolated and flattened (fig. 1.4*C*). Anuclear and binuclear superficial cells begin to appear. Superficial cells with a perinuclear halo are increasing in number.

4. *Ovulatory phase* (5–15 days, average 10 days). The ovulatory phase is characterized by an abundance (as high as 96% of all cellular elements) of superficial epithelial cells and an absence of mucus and leukocytes, unless their number is unusually high in the preceding phases (fig. 1.4*D, E*). Superficial cells, with brownish tan staining cytoplasm and numerous granules, gradually increase and reach their maximum number near the end of the phase (fig. 1.5*A*). There is also an increase in anuclear and binuclear superficial cells at this time. In one-fourth of the cases, intermenstrual bleeding (the "Hartman sign") occurs approximately 3 to 4 days preceding deturgescence of the sex skin and during the last few days of the ovulatory phase (fig. 1.5*A, B*).

The occurrence of ovulation is detected by the sudden decrease in superficial cells with brownish tan granular cytoplasm and in many instances intermenstrual bleeding (figs. 1.4*E;* 1.5*A*). It is significant to note that conception occurs most often (48% of the time) when mating occurs on the third day preceding deturgescence—at approximately the same time as intermenstrual bleeding (Hendrickx and Kraemer 1969). Smears taken after mating show a discoloration of the cells and a rather dirty appearance.

5. *Postovulatory phase* (3–8 days, average 6 days). The postovulatory phase is marked by the return of leukocytes and mucus as well as a clumping of cells (fig. 1.4*F*). Curling and folding of the cells also occur but are not as evident as the clumping (fig. 1.5*B, C, D*). Placard or rosette arrangements of cells are also quite common. The percentage of karyopyknotic cells remains high through the first few days and gradually diminishes near the end.

Almost all cellular elements are present in a typical postovulatory smear, particularly during the last few days. All types of superficial cells are represented,

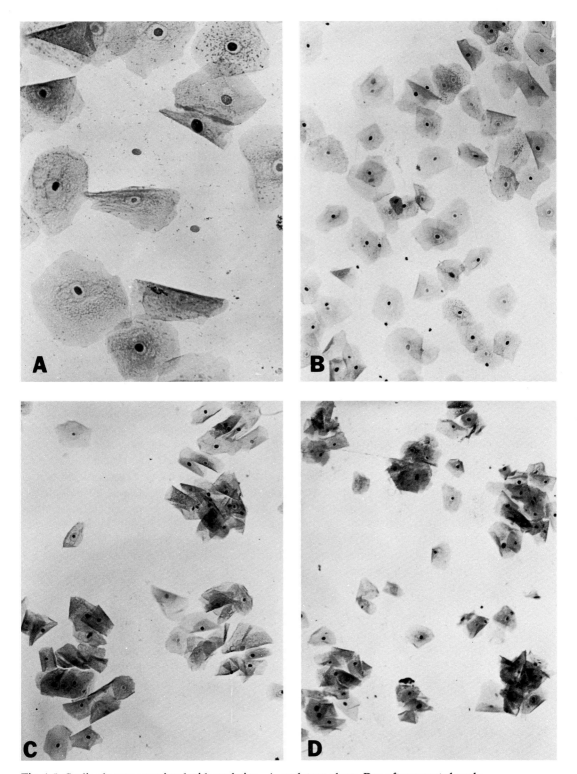

Fig. 1.5. Cyclic changes associated with ovulation. *A*, ovulatory phase. Day after expected ovulation. The Hartman sign or intermenstrual bleeding is indicated by the presence of erythrocytes in the vaginal smear. (×360) *B*, last day of the ovulatory phase. The cells are spread out and evenly dispersed. Compare to *C*. (×160) *C*, first day of the postovulatory phase. The cells are clumped together and the edges are curled. Compare to *B*. The percentage of pyknotic nuclei remains high. (×160) *D*, last day of the postovulatory phase. Clumping and curling of the cells are more pronounced and there is a decrease in pyknotic nuclei. (×140)

including degenerative cells, but those with the granular cytoplasm have decreased in number. The parabasal cells do not appear until the onset of the luteal phase.

6. *Luteal phase* (4–8 days, average 6 days). The luteal phase is marked by a drop in the karyopyknotic index and an increase in leukocytes and mucus (fig. 1.4*G, H*). There is a gradual diminution in the number of all superficial and upper-intermediate and parabasal cells. Thus, the cells are smaller overall. Some of the parabasal and basal cells take on the eosinophilic stain. Crowding of the cells is common. Placard and rosette formations occur occasionally. Wrinkling of the cell membrane and a loss of cell boundaries also occur.

Overall, the smear appears to be thinner but takes on a dirty background appearance. Bleeding occurs frequently during this phase. In approximately 40% of the cases, bleeding occurred between 6 and 9 days after the expected time of ovulation. It is possible that this luteal bleeding is associated with implantation. The cells become more randomly arranged in the later phases of the cycle as the mucus lessens. Well-preserved naked nuclei from intermediate cells are seen during this phase.

We have observed a slight rise in the cornification curve during the luteal phase which seemingly corresponds to the second peak in the human but it is not as extensive as that described for the monkey (De Allende, Shorr, and Hartman 1943).

7. *Premenstrual phase* (1–3 days, average 2 days). This phase is typified by the random arrangement of cells, a lessening of mucus, and a slight increase of superficial cells (fig. 1.4*I*). All cell types are present but they are fewer in number and tend to be smaller, suggesting that they are less mature and arise from a lower level of the vaginal epithelium. Bacteria and debris are more plentiful than in the previous phase. The presence of the premenstrual phase is dependent upon the previous luteal phase and the following menstrual phase. In other words, the existence of a premenstrual phase is determined in retrospect. If the mucus is thick and tends to obscure the cellular elements, then 3 or 4 days before the menstrual period the mucus liquifies and the slide is very neat grossly. Often there is a prolongation of the luteal phase to the onset of menstruation, and the premenstrual phase then is not clearly delineated. Of the 7 phases described, the postovulatory and premenstrual phases are the most transitory and the most apt to be overlooked. Figure 1.6 shows cellular elements that occur in the normal smear and figure 1.7 demonstrates the percentage of cell types which appeared in a normal 34-day cycle for animal A-292.

IV. Perineal Cycle

Visible cyclic changes occur in the sex skin or perineum which correlate with the menstrual cycle. These changes form the perineal cycle, which is divided into 2 phases each having 2 stages.

A. *Turgescent Phase*

1. *Initial turgescent stage* (average 4 days; fig. 1.8*A, B*). This stage begins when the perineal area starts to swell with a decrease in wrinkling of the skin that changes in color from dull pink to pinkish red. It terminates when the swelling reaches a plateau approximately 4 to 8 days after turgescence is first apparent. At that time the perineum is a deep red color and is so swollen that only small wrinkles are visible. The swelling surrounds both the anus and the vaginal vestibule. The clitoral lobes remain a light color.

2. *Maximum turgescent stage* (average 13 days; fig. 1–8*C*). This stage follows

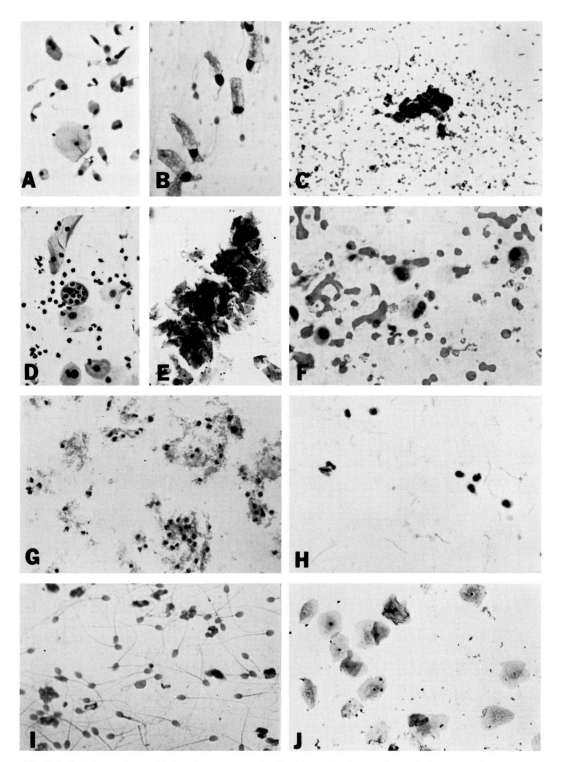

Fig. 1.6. Cellular and noncellular elements associated with cyclic changes in vaginal smears. *A*, endocervical cells in an ovulatory phase smear. Note that the shape varies from elongate to oval. (×175) *B*, endocervical cells in an ovulatory phase smear. Note the elongate shape, vacuolated cytoplasm and the nuclei in an eccentric position. (×440) *C*, sheet of endocervical cells in a menstrual phase smear. Note the hexagonal shape and central position of the nuclei. Histiocytes also are present. (×165) *D*, leukocytes engulfed by parabasal cells. Luteal phase. (×90) *E*, placard formation in an early luteal phase smear. Placard, or rosette, formation is indicative ∴ a progesterone effect. (×90) *F*, histiocytes in a menstrual phase smear. Note the foamy cytoplasm, indistinct cellular borders and variations in nuclear shape. (×525) *G*, cytolysis. The cytoplasm of the epithelial cells is frayed and fragmented. Note the free nuclei. (×175) *H*, bacillus vaginalis (Doderlein). Both long and short forms are present. (×440) *I*, spermatozoa in an ovulatory phase smear, 12 hours after mating. (×440) *J*, ovulatory phase smear, 1 day after mating. Note the dirty-dry appearance. Cells are relatively scarce. (×140)

initial turgescence and is the period when the perineum is swollen maximally. The skin of the perineum is fully distended and attains its deepest and most intense color, a bright red. The entire perineum is devoid of wrinkles and has a smooth, shiny appearance.

Gillman and Gilbert (1946), on the basis of perineal measurements, indicated that between 24 and 72 hours before the perineum achieves its maximum size a sudden decrease occurs in all the measurements for about 24 hours. According to them the perineum reaches its maximum dimensions within the next 2 or 3 days. Visual observations, which we used, are not conducive to such an assessment, but

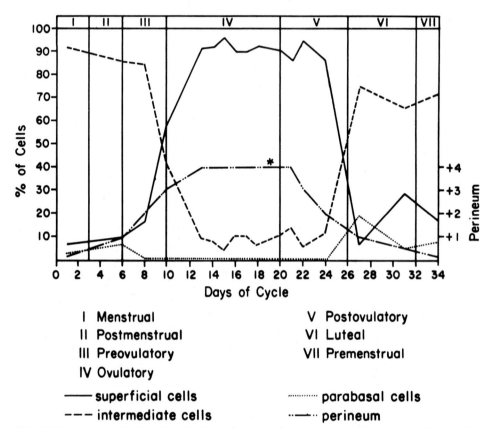

I   Menstrual   V   Postovulatory
II Postmenstrual   VI Luteal
III Preovulatory   VII Premenstrual
IV Ovulatory

——— superficial cells   ·········· parabasal cells
– – – intermediate cells   ·—·· perineum

Fig. 1.7. Percentage of epithelial cells found in the vaginal smear at various phases of the cycle. Note the correlation between the end of the ovulatory phase and the expected day of ovulation. * Expected time of ovulation.

in a small number of cases there was an apparent decrease in size followed by an increase. The increase in size is associated with ovulation as will be discussed in the section on optimal mating time. The size of the perineum during the maximum turgescent stage varies considerably among individuals.

B. *Deturgescent Phase*

1. *Initial deturgescent stage* (average 5 days; fig. 1.8*D, E*). This stage begins with a loss of color, a decrease in the size of the swelling, and a corresponding increase in wrinkles. The color changes from a bright to a dull red with overtones of gray and white. The first indication of changes in color and turgidity is in the clitoris followed by similar changes at the edge of the perineum which spread medially. Although the initial changes are rather subtle, a trained observer can

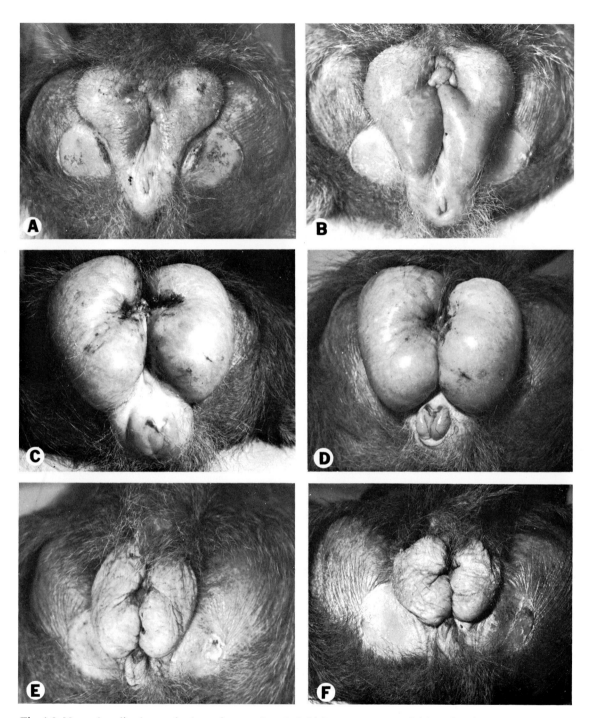

Fig. 1.8. Normal cyclic changes in the perineum. *A*, early initial turgescent stage, 2d day of cycle. *B*, late initial turgescent stage, 5th day of cycle. *C*, maximum turgescent stage, 16th day of cycle, 1 day after the time of anticipated ovulation. *D*, early initial deturgescent stage, 18th day of cycle. Note the marked decrease in size of the clitoral lobe. *E*, late initial deturgescent stage, 20th day of cycle. *F*, quiescent stage, 22d day of cycle.

detect them within an 8-hour interval. A further loss in color and turgidity follows and the sex skin appears flaccid in sharp contrast to the turgid appearance of the previous stage. The initial deturgescent stage varies in length from 2 to 14 days.

2. *Quiescent stage* (average 12 days; fig. 1.8F). During the quiescent stage the perineum is of minimal size. The labia and clitoris have many wrinkles of varying size with an overall pinkish red color. The epithelial surface of the perineum is dull compared to the bright and shiny surface prevalent during maximum turgescence. In some animals it begins to slough during the deturgescent stage, continues through the quiescent stage, and is usually completely shed by the beginning of the subsequent turgescent phase. It is not uncommon, however, to see remnants of the dried epithelium loosely attached at the onset of maximum turgescence. The length of the quiescent stage varies in direct proportion to the size of the perineum at maximum turgescence and lasts from 5 to 15 days.

In the pregnant animal the perineum remains at rest throughout the gestation period but the color changes slightly to a deeper red while the buttocks, the hairless areas located lateral to the labia, also have a lusty red color. The surface of the buttocks tends to become smooth and shiny. The color of the buttocks and the complete deturgescence of the perineum are quite reliable indicators of pregnancy by the 30th day of gestation, although the characteristic color may appear in some animals as early as the 14th day of gestation.

V. Cycle Length

Menstrual cycle length is determined by counting from the first day of overt menstrual flow (day 1) of one cycle up to the onset of bleeding in the following cycle. Perineal cycle length is determined by counting from the first day of perineal turgescence (day 1) of one cycle up to the onset of perineal turgescence in the following cycle. The average length of the menstrual cycle and the perineal cycle for *P. cynocephalus* in the Southwest Foundation colony is 33 days. This is quite similar to that presented by Zuckerman (1930, 1937) and Zuckerman and Parkes (1932) for *P. anubis* and *P. cynocephalus,* and is also closely correlated with *P. hamadryas* and *P. ursinus* (Gillman and Gilbert 1946) (see table 1.1). In approximately 16% of the cycles, overt menstruation does not occur in successive cycles, making it difficult to utilize this as an end point without resorting to vaginal smears. Although vaginal smears make it possible to detect menstruation in approximately

TABLE 1.1

CYCLE LENGTH IN DIFFERENT SPECIES OF BABOONS

	No. of Cycles	No. of Animals	Range (Days)	Mode (Days)	Mean ± SD (Days)
P. cynocephalus					
Menstrual	96	32	19–43	31	33.16± 3.74
Perineal	223	32	20–66	31	33.05± 5.28
*P. porcarius					
All	507	34	17–238	35	39.63±16.20
Selected	404	. . .	29–42	35	35.61± 3.20
*P. hamadryas					
1st series	72	10	22–46	33	31.4 ± 5.18
2d series	55	5	25–84	33	36.2 ±11.3
*P. cynocephalus	32	2	25–41	31–32	33.3 ± 3.48
*P. anubis	20	1	28–38	35	34.75± 2.38

* Data for these species from Eckstein and Zuckerman 1956, table 2, p. 344.

95% of the cycles, it is less convenient to measure than the changes that occur in the perineum.

Menstruation begins approximately 2 days before the onset of perineal turgescence and lasts about 3 days in *P. cynocephalus* (fig. 1.9). In *P. hamadryas* it occurs 3 to 4 days preceding turgescence and in *P. ursinus* turgescence begins on the 1st, 2d, or 3d day of menstruation in 82% of the cycles (Zuckerman and Parkes 1932; Gillman and Gilbert 1946). The turgescent and deturgescent phases of the cycle described previously approximate the follicular and luteal ovarian phases, respectively. The turgescent phase is 17 days and the deturgescent phase is 16 days for *P. cynocephalus*. The lengths of the turgescent and deturgescent phases are similar to the follicular and luteal phases described by Zuckerman (1937) for *P. hamadryas*. Gillman and Gilbert defined the turgescent phase as beginning with menstruation and ending with the onset of deturgescence in *P. ursinus* (table 1.2).

Because the major source of baboons for experimental purposes is from the wilds of East Africa it is of practical importance to determine what effect the rigors of capture and caging have on cycle length and fecundity. A period of transient

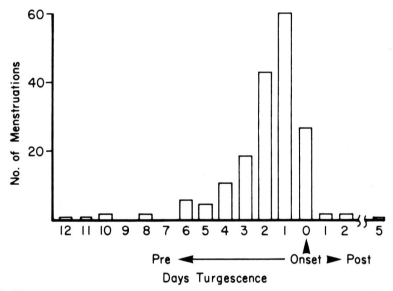

Fig. 1.9. The occurrence of menses relative to the onset of the perineal turgescence. (From Hendrickx and Kraemer 1969.)

TABLE 1.2

THE TURGESCENT (FOLLICULAR) AND DETURGESCENT (LUTEAL) PHASES
OF THE CYCLE IN DIFFERENT SPECIES OF BABOON

	No. of Cycles	Phases	Range (Days)	Mode (Days)	Mean ± SD (Days)
P. cynocephalus	226	turgescent	10–36	18	17.04 ± 3.60
		deturgescent	5–47	15	15.96 ± 4.14
*P. porcarius	404	follicular	10–32	19	19.45 ± 3.55
		luteal	6–25	17	16.07 ± 2.77
*P. hamadryas					
1st series	72	follicular	8–25	15–19	17.1 ± 3.49
		luteal	10–21	14	15.1 ± 2.47
2d series	55	follicular	12–69	17	21.8 ± 10.1
		luteal	9–22	15	14.4 ± 1.88

* Data for these species from Eckstein and Zuckerman 1956, table 3, p. 345.

infertility, i.e., irregular cycles and very low conception rate, is observed in baboons recently transported from Africa. Cycle length varies from 20 to 110 days for the first 6 months but then stabilizes quickly. By the end of 1 year in captivity the length ranges from 25 to 45 days with an average of 33 days. Figure 1.10 illustrates the range and mean of cycle length for 58 baboons for the first 13 months in captivity. All 58 females began cycling by the eighth month and 30% of them conceived within that period (fig. 1.11). Approximately 10% of the females conceived within 6 months after arrival (fig. 1.12). Forty percent conceived between 6 to 12 months after arrival. The remainder of the initial conceptions oc-

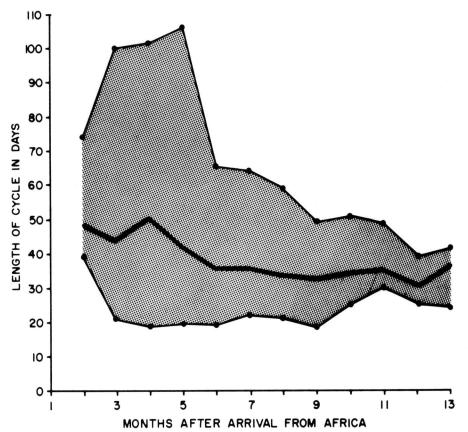

Fig. 1.10. The effect of adaptation to captivity on cycle duration (N = 58). (From Hendrickx and Kriewaldt 1967.)

curred between 12 and 24 months after arrival. Approximately 15% of them did not conceive but this is somewhat related to the size, weight, and age of the trapped animal. Animals that conceive within 6 months after arrival from Africa have a much shorter period of secondary amenorrhea than those that conceive after 6 months (table 1.3). The length of the menstrual cycle is within the normal range for all animals, but those that conceive within 6 months have approximately 3 cycles before conception whereas those that conceive after 1 year average approximately 8 cycles before conception.

VI. Optimal Mating Time

In the preceding section the cyclic changes of the perineum and vaginal cytology are presented. In this section evidence will be presented which indicates that the period of maximum fertility is best defined on the basis of the perineal cycle.

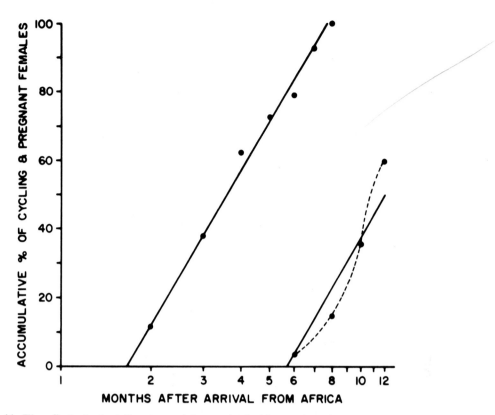

Fig. 1.11. The effect of adaptation to captivity on the incidence of cycling ($N = 58$). (From Hendrickx and Kriewaldt 1967.)

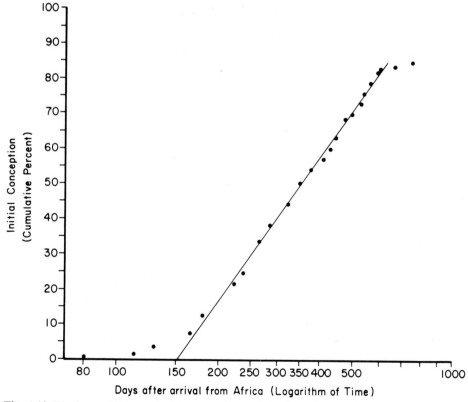

Fig. 1.12. The interval between initial conceptions and the time of arrival from Africa ($N = 133$). (From Kriewaldt and Hendrickx 1968.)

The objective of the experiment from which this evidence is obtained is to establish a temporal and morphological relationship for the embryos. Therefore, mating periods are as short as circumstances permit. For the most part a fertile male baboon is introduced into the female's cage for an 8-hour period between 8:00 A.M. and 5:00 P.M. of a regular working day. Although some matings are scheduled for overnight (12 hours) the shortest recorded mating time for a conception is 2 minutes. These relatively short periods of exposure of the female to the male that occur once during the cycle are considered single matings.

Zuckerman (1930, 1937) and Zuckerman and Parkes (1932) conclude that ovulation coincides with sex skin deturgescence. Gillman and Gilbert (1946) suggest on limited evidence that ovulation may precede deturgescence by at least 2 or 3 days. Consequently mating is scheduled as close to 3 days preceding deturgescence as can be predicted on the basis of previous cycles. However, if the

TABLE 1.3

DURATION OF SECONDARY AMENORRHEA, AVERAGE MENSTRUAL
CYCLE LENGTH, AND THE AVERAGE NUMBER OF CYCLES
BEFORE INITIAL CONCEPTION

	Secondary Amenorrhea (Mean Days)	Average Menstrual Cycle Length (Days)	Average No. of Cycles before Conception
Early conceivers <180 days (N=13)	46.2	44.6	2.6
Average conceivers 180–360 days (N=56)	97.4	36.7	4.4
Late conceivers >360 days (N=34)	98.7	43.5	8.4

SOURCE: Kriewaldt and Hendrickx 1968, p. 367.

perineum has not commenced deturgescence by the 3d or 4th day after mating, the male is returned to the female's cage for another 2- to 12-hour period until evidence of perineal deturgescence is observed. The mating procedure in which the female is exposed more than once during the cycle is considered a continuous mating.

The occurrences of 195 fertile and nonfertile single matings on the various days preceding deturgescence are shown in figure 1.13. The interval of 2 to 4 days preceding deturgescence is predictable 60% of the time, and day 3 preceding deturgescence is predictable 25% of the time. The conception percentages for the continuous matings are in figure 1.14. Although there is no clear-cut statistical difference between conception rates when matings occur on days 3 through 7 preceding deturgescence, the highest conception rate (48%) is observed when mating occurs on day 3 preceding deturgescence. This compares favorably with the conception rate (41%) of continuous matings, which occur throughout the last half of the turgescent phase (fig. 1.15). This observation is strengthened somewhat by the fact that the conception rate for continuous matings from day 4 or less preceding deturgescence is twice that obtained where matings are restricted to the period of 5 or more days preceding deturgescence (fig. 1.15).

The correlation of chronological age and morphological development is discussed under "Age and Size" in chapters 3 through 8. It can be stated at this point, however, that based on conceptions following single matings, ovulation occurs most often 3 or more days preceding deturgescence. In several instances where an embryo in the pronuclear or syngamy stage of development was desired, a direct examination of the ovaries was made by laparotomy on the day preceding the estimated time of ovulation (day 4). In these cases, ovulation had not occurred, but when the laparotomy was repeated, ovulation had occurred and deturgescence commenced 2 or 3 days later.

The days preceding deturgescence have more predictive value for determining optimal mating time than the cycle day. Figure 1.16 shows that fertile matings occur on all days from 9 through 20 of the perineal cycle with the highest conception rate (6 of 8 matings, 75%) occurring on cycle day 17. Efforts to predict optimal mating time by relating mating to the midpoint of the perineal cycle are

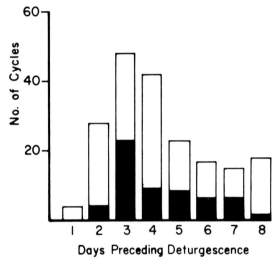

Fig. 1.13. Fertile (*solid columns*) and nonfertile (*open columns*) single matings on days preceding deturgescence. (From Hendrickx and Kraemer 1969.)

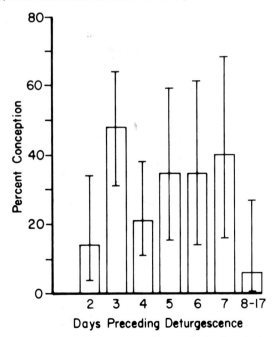

Fig. 1.14. Conception percentages and 95% confidence limits for single matings on days preceding deturgescence. (From Hendrickx and Kraemer 1969.)

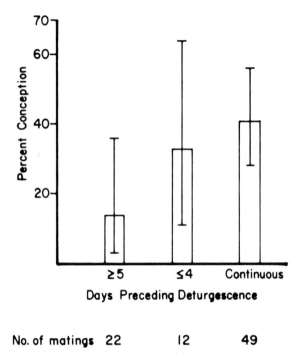

No. of matings   22   12   49

Fig. 1.15. Conception percentage and 95% confidence limits for multiple matings during the middle (day 5 preceding deturgescence and earlier), late (days 1 through 4 preceding deturgescence), and maximum turgescent stages. (From Hendrickx and Kraemer 1969.)

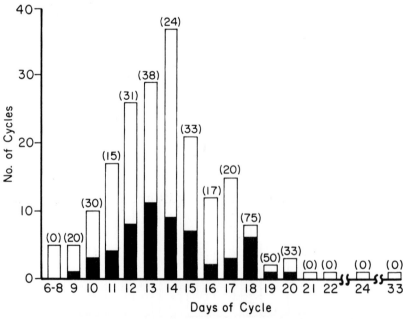

Fig. 1.16. Fertile (*solid columns*) and nonfertile (*open columns*) single matings on days of perineal cycle. Figure in parentheses gives percent of conception on that day. (From Hendrickx and Kraemer 1969.)

24

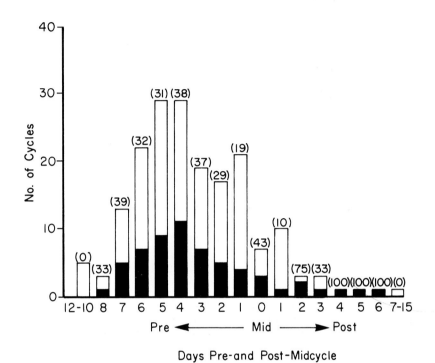

Fig. 1.17. Fertile (*solid columns*) and nonfertile (*open columns*) single matings on days pre- and post-midcycle. Figure in parentheses gives percent of conception on that day. (From Hendrickx and Kraemer 1969.)

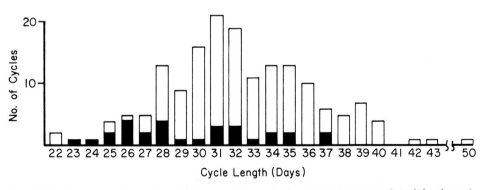

Fig. 1.18. Frequency of cycle lengths in mated (*open columns*) and nonmated (*solid columns*) cycles. (From Hendrickx and Kraemer 1969.)

♀ A167 (average cycle length 28 days)

	JAN	FEB	MAR	APR	MAY	JUNE	JULY	AUG	SEPT	OCT	NOV	DEC
1964								38	Pg			
1965		37	28 Pg		30	30	Pg	27	26	30	25	29
1966	27	28	25 EN	28	25	EN	30	26	EN	26	31	25
1967	26 26	28	EN	23	Pg	28	28	26	29	23	37	Pg

♀ A336 (average cycle length 31 days)

	JAN	FEB	MAR	APR	MAY	JUNE	JULY	AUG	SEPT	OCT	NOV	DEC
1964								31	Pg	34		Pg
1965	37	Pg			29	35	Pg		30	34	25	36
1966	30	Pg		Pg					30	27	30	
1967	32	30	30	EN	29	34	29	EN				

♀ A95 (average cycle length 37 days)

	JAN	FEB	MAR	APR	MAY	JUNE	JULY	AUG	SEPT	OCT	NOV	DEC
1964									52		38	38
1965		38	39	Pg		40	Pg	39	36		Pg	
1966	Pg		40	35	Pg							42
1967	34	Pg						27	19 EN			

Fig. 1.19. Frequency of conception and variation in cycle length in females with cycles of different average length. Cycles calculated per half-month period. *Pg*, pregnant; *EN*, removal of the endometrium.

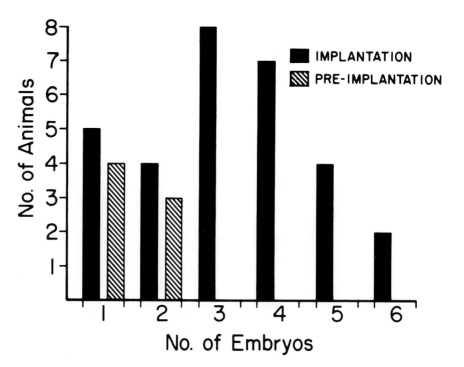

Fig. 1.20. The number of embryos collected from 32 fertile females. N = 117 embryos, average 3.7 embryos/female.

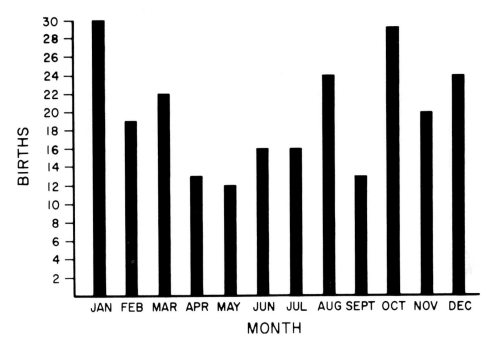

Fig. 1.21. Occurrence of births in captive baboons for a 2-year period. (From Kriewaldt and Hendrickx 1968.)

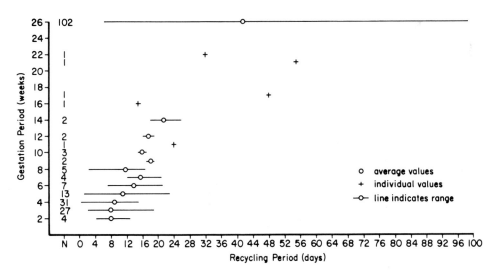

Fig. 1.22. Effect of surgical termination of pregnancy (embryotomy), at various times during gestation, on recycling time. (From Kriewaldt and Hendrickx 1968.)

27

also less reliable than using the time of deturgescence (fig. 1.17). Conceptions occur on days ranging from 8 days before to 6 days after midcycle.

The significance of the above data is that to accurately and consistently predict the optimal mating time in baboons one must consider each female as an individual. After recording and studying 3 or 4 cycles it is then possible to estimate the optimal mating time with reasonable success. In baboons, at least, cycle length within individuals is quite constant but quite variable between individuals. Figure 1.18 shows the variation in cycle lengths between individuals. For this reason mating is scheduled for the 3d day preceding deturgescence on an individual basis instead of a predetermined day of the cycle. If one could select females with the same cycle length then the latter would be possible. Cycle length, however, has relatively little bearing on fecundity. Figure 1.19 shows the cycle lengths and conceptions of

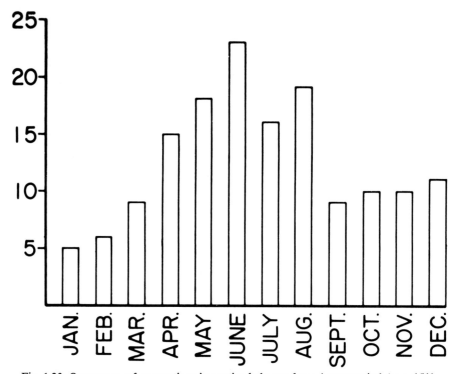

Fig. 1.23. Occurrence of conceptions in captive baboons for a 4-year period (N = 151).

3 baboons from which embryos were collected. Animal A-167 had the shortest cycles, A-95 the longest, and the cycles of A-336 were the same length as the mode (31 days). The number of embryos collected per female from a group of 32 proven breeders are shown in figure 1.20.

VII. Birth Rates

The baboon is monotocous but occasionally twinning occurs. The incidence of twinning varies between colonies. Four sets of twins in 730 deliveries have been recorded in the Southwest Foundation baboon colony (Hendrickx, Houston, and Kraemer 1968), and two sets of twins in 837 deliveries have been recorded at the Sukhumi colony (Lapin and Yakovleva 1963).

The baboon does not have distinct seasonal birth peaks although the majority of births occur during the fall and winter months (fig. 1.21). This means that the majority of conceptions occur in the late spring and early summer months because the gestation period is approximately 175 days. The average recycling period, the time from birth until the first onset of turgescence, is approximately 44 days, with a range of 21–67 days.

Surgical intervention, i.e., removal of the embryo, in early pregnancy tends to shorten the recycling period, with a consequent slight alteration of birth peaks (fig. 1.22). However, there is still a higher incidence of conceptions in the spring and summer months (fig. 1.23).

References

De Allende, I. L. C.; E. Shorr; and C. G. Hartman 1943. A comparative study of the vaginal smear cycle of the rhesus monkey and the human. *Contrib. Embryol., Carneg. Inst.* 31:1–26.

Eckstein, P., and S. Zuckerman 1956. The oestrous cycle in the mammalia. In F. H. Marshall, *Physiology of Reproduction,* ed. A. S. Parkes, 3d ed., vol. I, part I, pp. 226–396. London: Longmans Green.

Gillman, J., and C. Gilbert 1946. The reproductive cycle of the chacma baboon (*Papio ursinus*) with special reference to the problems of menstrual irregularities as assessed by the behavior of the sex skin. *S. Afr. J. Med. Sci.,* suppl., 11:1–54.

Gold, J. J., ed. 1968. *Textbook of gynecologic endocrinology.* New York: Harper & Row.

Graham, R. M. 1963. *The cytologic diagnosis of cancer.* 2d ed. Philadelphia and London: W. B. Saunders.

Gray, H. 1959. *Anatomy of the human body.* Edited by C. M. Goss. 28th ed. Philadelphia: Lea & Febiger.

Hartman, C. G. 1961. *The anatomy of the rhesus monkey,* Macaca mulatta. New York: Hafner.

Hendrickx, A. G. 1967. The menstrual cycle of the baboon as determined by the vaginal smear, vaginal biopsy, and perineal swelling. In *The Baboon in Medical Research,* ed. H. Vagtborg, 2:437–59. Austin: University of Texas Press.

Hendrickx, A. G.; M. L. Houston; and D. C. Kraemer 1968. Observations on twin baboon embryos (*Papio* sp.). *Anat. Rec.* 160:181–86.

Hendrickx, A. G., and D. C. Kraemer 1969. Observations of the menstrual cycle, optimal mating time, and preimplantation embryos of the baboon. *J. Reprod. Fert.,* suppl., 6:119–28.

Hendrickx, A. G., and F. H. Kriewaldt 1967. Observations on a controlled breeding colony of baboons. In *The Baboon in Medical Research,* ed. H. Vagtborg, 2:69–83. Austin: University of Texas Press.

Hill, W. C. O. 1967. Taxonomy of the baboon. In *The Baboon in Medical Research,* ed. H. Vagtborg 2:3–11. Austin: University of Texas Press.

Koss, L. G. 1968. Diagnostic cytology and its histopathologic bases. 2d ed. Philadelphia and Toronto: J. B. Lippincott.

Kriewaldt, F. H., and A. G. Hendrickx 1968. Reproductive parameters of the baboon. *Lab. Anim. Care* 18:361–70.

Lapin, B. A., and L. A. Yakovleva 1963. *Comparative pathology in monkeys.* Translated by U.S. Joint Publ. Res. Service. Springfield, Ill.: Charles C. Thomas.

Maples, W. R. 1967. Classification of the Kenya baboon. Ph.D. dissertation, University of Texas.

Papanicolaou, G. N. 1963. *Atlas of exfoliative cytology.* Cambridge: Harvard University Press.

Smolka, H., and H.-J. Soost 1965. *An outline and atlas of gynaecological cytodiagnosis.* London: Edward Arnold.

Zuckerman, S. 1930. The menstrual cycle of the primates: I. General nature and homology. *Proc. Zool. Soc. Lond.*, no. 45, pp. 691–754.

———— 1937. The duration and phases of the menstrual cycle in primates. *Proc. Zool. Soc. Lond.*, ser. A, part 3, pp. 315–29.

Zuckerman, S., and A. S. Parkes 1932. The menstrual cycle of the primates: V. The cycle of the baboon. *Proc. Zool. Soc. Lond.*, pp. 139–91.

2

Methods

Andrew G. Hendrickx/Duane C. Kraemer

I. Management and Breeding Techniques

A. *Housing and Handling*

The embryology breeding colony at the Southwest Foundation for Research and Education is kept in an indoor breeding unit which consists of a combined record and treatment room, a storage and cage-cleaning area, and a caging area which contains room for 68 to 72 individual hanging cages. The caging area is subdivided into 3 bays each containing 4 rows of 6 cages per row. The two alleys which run the length of the caging area are flanked by cages on each side. The cages open onto the alley and are accessible from the rear by a narrow walkway. The animals are arranged so that 1 male servicing 11 females is in visual, auditory, and olfactory contact with them at all times (fig. 2.1).

At the selected time for mating, the male is placed in a portable transfer cage and brought to the female's cage. This procedure is used rather than delivering the female to the male, which is common practice for other nonhuman primates, for two reasons: (1) it is easier to train 1 male than 11 females to enter and leave a transfer cage, and (2) after mating, the male is more willing to leave the female's cage than he is to permit her to leave his cage. In the earlier part of the study the females were taken to the male but this often resulted in a struggle to separate the two with the male as the main objector. Generally, females are more receptive to males kept in the immediate area than those brought in from the adjoining bay. However, the opposite reaction has been observed.

B. *Menstrual Cycle Recording*

The cyclic changes of the menstrual cycle, including the perineal changes, are presented in chapter 1. The perineum is examined and recorded daily on a breeding chart kept for each female using a numerical grade point system of 0 to 5 (fig. 2.2). A designation of 0 indicates perineal rest, a 1, 2, or 3 indicates either perineal turgescence or deturgescence, depending on its relation to 4, the maximum turgescent stage. A designation of 5 indicates a further increase in turgescence, which occasionally occurs late in the maximum turgescent stage. A descending arrow indicates the start of deturgescence. The examination is usually done by two technicians, each observing the perineum from opposite sides of the cage. The changes in its size and color and the appearance of menstrual blood are then recorded on

31

the breeding chart. Clinical data, including administration of medication, sedatives, anesthesia, and surgery, are recorded on a clinical record sheet kept for each animal.

C. *Mating*

Mating is indicated on the breeding record at each occurrence by a solid black square with the male's number placed vertically. The time of entrance and exit of the male from the female's cage is recorded if it is longer or shorter than an approximate 8-hour period. Mating is scheduled for one 8-hour period during the cycle, on the 3d day preceding deturgescence (chap. 1). In predicting this optimal mating day, the length of the turgescent phase of several previous cycles must be known. It is also helpful to know the length of the entire cycle and the time of menses but they are of secondary importance to knowing the 1st day of deturgescence. If the

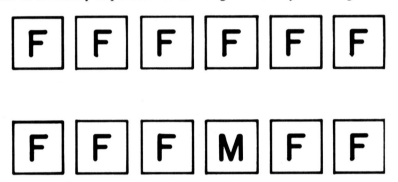

Fig. 2.1. Cage arrangement for breeding baboons in individual cages.

BABOON MENSTRUAL CYCLE RECORD

Animal No. _____   Year _____

Date	6/7	6/8	6/9	6/10	6/11	6/12	6/13	6/14	6/15	6/16	6/17	6/18	6/19	6/20	6/21	6/22	6/23	6/24	6/25	6/26	6/27	6/28	6/29	6/30	7/1	7/2	7/3	7/4	7/5	7/6	7/7	7/8	7/9	7/10	7/11
Cycle Day	31	32	33	34	1	2	3	4	5	6	7	8	9	10	11	12	13	14	15	16	17	18	19	20	21	22	23	24	25	26	27	28	29	30	31
+5																																			
+4																																			
+3																																			
+2																																			
+1																																			
0																																			
Remarks						M	M																	A503											

Date	7/12	7/13	7/14	7/15	7/16	7/17	7/18	7/19	7/20	7/21	7/22	7/23	7/24	7/25	7/26	7/27	7/28	7/29	7/30	7/31	8/1	8/2	8/3	8/4	8/5	8/6	8/7	8/8	8/9	8/10	8/11	8/12	8/13	8/14	8/15
Cycle Day	32	33	34	35	36	37	38	39	40	41	42	43	44	45	46	47	48	49	50	51	52	53	54	55	56	57	58	59	60	61	62	63	64	1	2
+5																																			
+4																	IA and EFA (36 d.)																		
+3																																			
+2																																			
+1												A68-161																							
0																																			
Remarks																		CS CL																	

Fig. 2.2. Menstrual cycle record used in baboon breeding program. All essential data are recorded on the record sheet. ■ = day of mating; *M*, menses; *CS*, cesarean section; *CL*, corpus luteum removed at the time of surgery; *IA*, insemination age; *EFA*, estimated fertilization age.

cycle is longer than the previous cycles, mating can be repeated every third day until deturgescence occurs.

Insemination or deposition of semen in the vagina is determined by one of the following methods: (1) observation during copulation, (2) presence of seminal fluid or coagulum on the female's external genitalia, or (3) presence of sperm in the vaginal smear. The latter method is the most reliable. In most instances the smear is examined immediately with a phase contrast microscope and then stained by the same procedure used for vaginal cytology studies. It is then reexamined and stored.

II. Determining Embryonic Age

Embryonic age is defined as the best estimate of the fertilization age of the embryo based on the following factors: (1) day of insemination, (2) day of deturgescence, and (3) optimal mating time. In considering these factors, three possible ages for an embryo may be determined. The possible ages of the embryo include insemination age, embryonic age (estimated fertilization age), and minimum age.

Insemination age is the maximum possible age for the embryo and is calculated by designating the day of insemination as day 0 in single matings. If matings are repeated, as is the case in unpredictably long cycles, the day of first insemination is considered 0. The insemination age is not calculated in colony matings. These mating procedures in which the female is exposed more than once during the cycle are considered continuous matings.

The estimated fertilization age is calculated by considering the 3d day preceding deturgescence as day 0 because this is when the highest percentage of conceptions (48%) occur (see chap. 1). This rule is applied in all cases except when single matings occur on or before the 5th day preceding deturgescence. In these instances, the estimated fertilization age is calculated by considering the 2d day after insemination as day 0. This is based on an undocumented assumption that by 2 days after insemination the sperm have lost their fertilizing capacity.

Minimum age is calculated by designating day 2 preceding deturgescence as day 0 of gestation since there are no conceptions when matings occur on the day immediately preceding deturgescence, suggesting that the ovum does not survive beyond this point in the cycle (chap. 1). An exception to this method of calculating minimum age is when single matings occur on day 5 or earlier preceding deturgescence (fig. 2.3D). In this instance minimum age is 2 days later than insemination age and is the same as estimated fertilization age (chap. 1). Unless a single mating occurs on day 2 preceding deturgescence, day 3 is the best estimate of fertilization age. When a single and/or initial mating occurs at the optimal mating time (day 3), the embryonic age (estimated fertilization age) is the same as insemination age.

The examples shown in figure 2.3 illustrate the possibilities that may occur in a breeding colony using this method. Figure 2.3A shows that when mating occurs at the optimal time, the 3d day preceding deturgescence, the estimated fertilization age and the insemination age are the same and that the minimum age is 1 day less. Figure 2.3B shows that if mating occurs on the 2d day preceding deturgescence all 3 ages are the same. This is the only instance when this occurs. In figure 2.3C all 3 ages are different. The 1st day of mating is the insemination age, day 3 preceding deturgescence is the estimated fertilization age, and day 2 preceding deturgescence, which is the 2d mating in this example, is the minimum age. Figure 2.3D indicates the method of calculating embryonic age when mating occurs earlier than the 5th day preceding deturgescence. The day of mating is the insemi-

EXAMPLES OF METHODS
FOR DETERMINING EMBRYONIC AGE

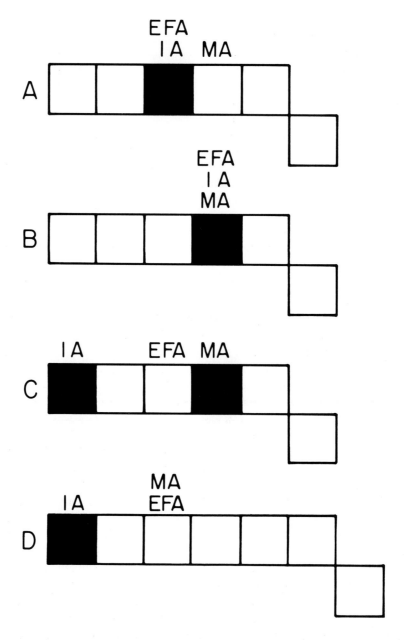

Fig. 2.3. Examples of methods for determining embryonic age. ■ = day of mating; *EFA*, estimated fertilization age; *IA*, insemination age; *MA*, minimum age. The day indicated by one of the above abbreviations is always considered day 0 for that particular age.

nation age, and the 2d day after insemination is the estimated fertilization age and the minimum age. In all instances the day the embryo is removed from the reproductive tract is considered as a day of development. The time of day at which surgery is performed varies only a maximum of 6 hours.

III. Collection and Processing of Embryos

Baboon embryos can be obtained either at autopsy, at hysterectomy, or by *in situ* surgical procedures. Surgical techniques which allow repeated use of the same female provide 3 distinct advantages: (1) animals become adjusted to the routine of cage confinement, thus exhibiting less cyclic fluctuation than unconditioned animals and providing more reliable data, (2) menstrual cycle data may be collected over long periods of time, and (3) the cost per embryo is significantly reduced. The total number of recovery attempts permitted on an individual female depends upon the stage of pregnancy involved. For example, when postimplantation stage embryos are obtained by cesarean section as many as 8 operations may be performed before sufficient adhesions result to severely affect fertility. On the other hand, when the uterine tubes and ovaries are handled in the flushing procedures necessary for *in situ* collection of preimplantation embryos, the formation of adhesions between the uterine tubes, ovaries, and uterus generally limits the number of collections to 4 or less.

A. *Preimplantation Embryos*

The preimplantation embryo may be located in either the uterine tube or the uterus. The collection procedure will vary slightly depending upon its expected location and the necessity of establishing its location. If the location of the embryo is uncertain or if the establishment of location is not important, the uterine tubes and uterus are flushed simultaneously (fig. 2.4). A glass speculum, 17 to 24 mm in diameter is inserted into the vagina and held firmly against the external cervical os. The uterus, uterine tubes, and ovaries are exposed by a midventral laparotomy. Fluted plastic catheters (Intramedic, PE160 or PE100, Clay Adams, N.Y.) are inserted approximately 3 mm into the infundibular ostia of the uterine tubes and secured by either a ligature or a Hagenbarth wound clip applicator. An 18 gauge, intravenous catheter (Jelco Labs, Raritan, N.J.) approximately 5 cm in length is inserted through the uterine wall, with the tip in the uterine lumen, and secured by a purse-string suture. Using a 30 cc syringe, the flushing fluid is forced through the uterus and out the cervix and uterine tubes. The fluid flowing through the cervix is collected in disposable petri dishes and that from the uterine tubes is collected in embryological watch glasses. After approximately 90 cc of fluid has been collected via the cervix, the cervix is clamped and fluid is then forced through the uterine tubes. Embryos obtained in the washings collected via the cervix are positively identifiable as having been located in the uterus at the time of collection. Those obtained in washings collected via the uterine tubes could have been located in either the uterus or uterine tube.

If one wishes to obtain an embryo with positive identification of its location in the uterine tube, the uterine tube wall can be punctured with a 25–30 gauge hypodermic needle so that fluid can be forced out the tubal cannula.

A dissecting microscope with magnification of 10–30 times is used to locate the embryos in the flushing fluid. When disposable petri dishes are used as the collecting vessels they are placed upon a glass slide which has been etched with lines approximately 1 cm apart to serve as guides to facilitate a thorough search of the entire dish.

The embryos are transferred from the washings, using a Pasteur pipette, into an embryological watch glass containing the fixative. If the embryos are to be stained as whole mounts, they are fixed in phosphate-buffered 10% formalin; if they are to be sectioned, they are fixed in 3–5% glutaraldehyde. After fixation the embryos are transferred to a depression slide and photographed using a Zeiss Ultraphot II and a Leitz phase contrast microscope. Comments are recorded regarding fertilization, number of cells, signs of degeneration, etc. At this time an outline drawing is made of the embryo to aid in later interpretation of the photographs. If the embryo is to be stored before further processing, it is transferred back to the fixative.

Whole mounts are prepared according to the method of Marston, Yanagimachi, Chang, and Hunt (1964) for mouse and golden hamster eggs. The embryos are rinsed in distilled water and dehydrated for 2–4 hours in freshly prepared 95% ethyl alcohol. They are then mounted on microscope slides and covered with a

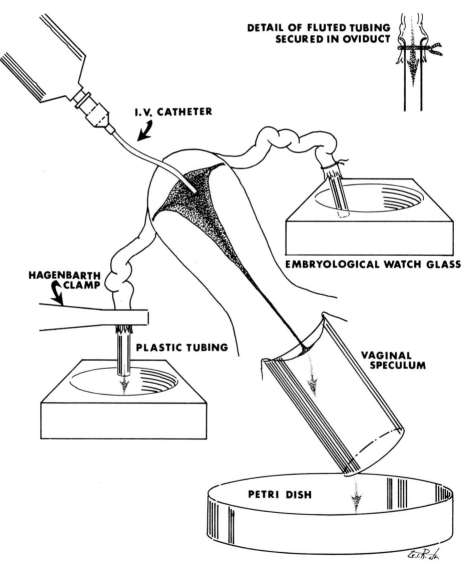

Fig. 2.4. Illustration of technique used to collect preimplantation baboon embryos.

coverslip supported by drops of 20 parts vaseline and 1 part paraffin wax. The space under the coverslip is then flooded with aceto-lacmoid (0.25% lacmoid in 45% acetic acid) until the nuclear elements are clearly defined. The excess stain is then removed by infiltrating under the coverslip with aceto-glycerol mounting medium (20% glycerol, 20% glacial acetic acid, and 60% distilled water). The edges of the coverslip are then sealed with Permount.

Embryos that are to be sectioned are rinsed in distilled water. Agar (1.7% aq) is heated on a hot plate until melted and several drops are placed in a 2 × 3 inch depression slide. The embryo is quickly transferred with a pipette from water to agar. Additional drops of agar are placed over the embryo. Orientation is achieved by moving a heated fine dissecting needle in a circular motion through the agar around the embryo. After the agar is hardened the block is trimmed to approximately 4 × 4 mm and placed in fresh fixative overnight.

The processing schedule is as follows: distilled water (several changes, total 30 min); 50%, 70%, 80%, and 95% alcohol (30 min each); absolute alcohol (2 changes, 30 min each); equal parts absolute alcohol and benzene (30 min); benzene (2 changes, 30 min each); paraplast (3 changes, 20 min each, under vacuum 5 or 10 lbs), followed by embedding in paraplast. The embryo is serially sectioned at 7μ and stained with Harris's hematoxylin and eosin.

B. *Postimplantation Embryos*

A ventral midline incision is made approximately 8.0 cm long and extending 3.0 cm from the anterior inlet of the pelvis. A Balfour retractor is used to expose the gravid uterus, which is delivered manually through the incision. The implantation site can be recognized as an increased area of myometrial vascularity with reasonable success as early as the 20th day of pregnancy. A sponge forceps is placed at the cervix uteri restricting the uterine blood supply. When possible the initial uterine incision is made on the side opposite the implantation site in the superficial layer of the myometrium, from the region anterior to the cervix uteri to the fundus. The initial incision is slowly enlarged until the endometrium is exposed in one small area. The incised edges of the myometrium are held in apposition with Allis tissue forceps while the incision is extended, by use of Strabismus scissors, through the intact layers of myometrium (fig. 2.5A). The incision site is kept free of blood by directing a steady stream of physiological saline at the point of incision. The endometrium which varies in thickness from 8 mm during the 3d week to 5mm during the 8th week of pregnancy is removed as an intact sac by separating the decidual layers from the basal layer with Strabismus scissors and a dental spatula. The connective tissue septae are cut with scissors and the friable blood vessels are detached by blunt dissection with the spatula. During the final stages of excision, digital pressure applied to the outside of the uterus at the implantation site aids in excision by partially everting the uterus and exposing the base of the vascular placental attachment. The endometrial sac containing the chorionic vesicle is excised anteriorly to the cervix uteri (fig. 2.5B). The uterine incision is closed using a modified Kushing suture in the myometrium and a simple continuous suture in the perimetrium (Hartman 1944; Claborn, Hendrickx, and Kriewaldt 1967).

When collecting the very early implantation stages it is beneficial to leave strips of myometrium on the endometrial sac. These muscle fibers provide rigidity to the small endometrial sac and also decrease the possibility of penetrating into the uterine cavity and destroying the implantation site. The collection of artifact-free embryos and placentae is the most difficult during this period of development, and it is sometimes necessary to resort to hysterectomy to get the necessary material. The most frequent artifact occurring in the early implantation period is a disruption of the placental surface, which destroys normal relationships. Also, maternal blood engulfs the embryo and fills the amniotic and vitelline cavity.

C. Complications

The major limiting factor in applying this technique is the development of adhesions between the incision site and omentum; between the uterus and adnexa; and between the ovary, uterine tube, and uterus and the broad ligament. Despite these factors as many as 8 enucleations have been performed on one animal in a 3-year period, with an average of 4 or 5 per animal. Figure 2.6 shows the number of cesarean sections and/or tubal-uterine flushings on 79 females.

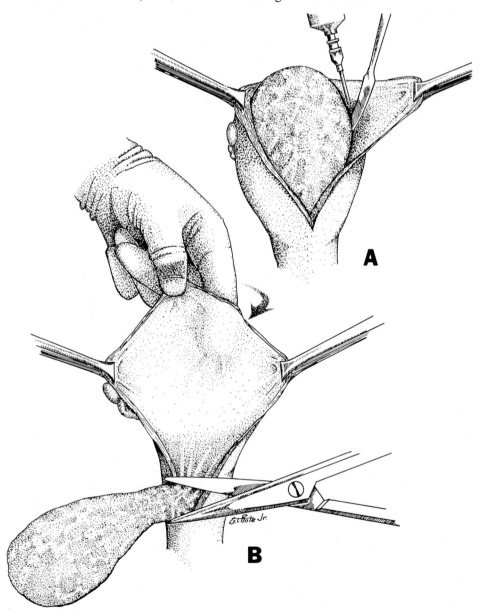

Fig. 2.5. Illustration of technique used to collect baboon embryos after implantation. *A*, a spatula is used to separate the endometrium (*EN*) from the myometrium (*MY*). *B*, the uterus is inverted (*arrow*) and the endometrial sac, containing the embryo, is cut at the cervix uteri.

D. *Dissection and Processing*

The endometrial sac is placed in a finger bowl immediately after removal and immersed in isotonic physiological saline. The *in vitro* endometrial sac is oriented similar to its *in situ* position in order to determine the site of attachment after dissection. Orientation is simplified by identifying the small tear in the endometrial sac at the initial incision site, or by the impression made by a hemostat on the cut edge of the cervix uteri. After a thorough rinsing to remove the blood from the surface, the endometrial sac is again immersed in saline. The dissection is begun by placing a small forceps on the cut edge of the cervical end of the endometrial sac and lifting it to expose the lumen. The endometrial sac is carefully opened by cutting through the thinner lateral walls until the implantation site is located (fig. 2.7). In the very young embryos implantation may be indicated by small hemorrhages on the endometrial surface or simply by a deep red "blushed" area. The implantation site of embryos 12 days old or older are relatively large and easily identified. After the implantation site is identified with certainty, the membranous chorion is exposed by cutting the sac on one side and across the

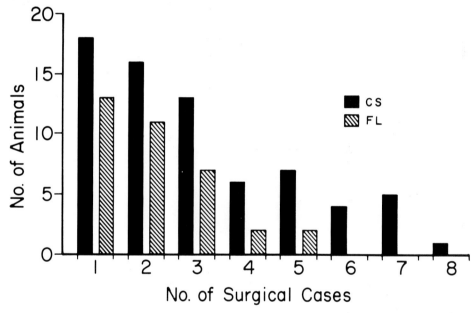

Fig. 2.6. The number of cesarean sections (*CS*) and uterine and tubal flushings (*FL*) attempted on 79 females. N = 105 surgical procedures.

fundic margin or by cutting away the entire abembryonic half of the endometrium. Complete exposure of the embryo within the endometrial sac, particularly in early implantation stages, is accomplished before fixation because it is easier and probably results in less artifact due to the greater elasticity of the unfixed tissue.

Fixation is accomplished with either Bouin's fluid, 10% buffered formalin, formol-saline, glutaraldehyde, or FAA (5 parts formalin, 5 parts acetic acid, and 90 parts 80% ethyl alcohol) by slowly adding the fixative to the saline solution until a concentration of about 25 to 50% is achieved. Bouin's fluid tends to coagulate the fluid of the chorion and the amnion, thus masking the presomite or early somite embryos and making it difficult to orient them properly for sectioning. Therefore, care should be taken to carefully wash away the fluids before fixation, or a different fixative should be selected. The diameter of the membranous chorion is measured and then dissected away in all embryos 15 days old or older. In embryos 25 days old or older the amnion is usually opened but not removed

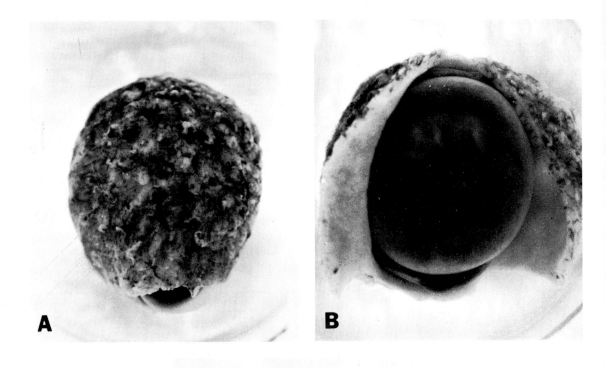

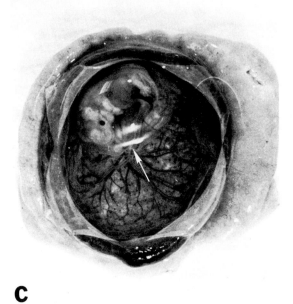

Fig. 2.7. Illustration of dissection procedure used in processing baboon embryos. *A,* undissected endometrial sac. The chorion protrudes at the point that the endometrium was cut from the cervix uteri. *B,* the endometrial sac is opened by making a longitudinal incision opposite the placenta, exposing the membranous chorion. Note the flaps of the abembryonic endometrium on both sides of the membranous chorion. *C,* the embryo, contained within the amnion, is exposed by making an incision through the membranous chorion and pushing it aside. The embryo is removed by cutting the umbilical cord at the edge of the amnion (*arrow*) or by removing the amnion first and then cutting the umbilical cord near the body. (×1.3)

until after fixation. Greatest length or crown-rump measurements are made immediately after adding the fixative and are considered unfixed measurements. The fixative is then serially replaced to make the solution full strength. Specimens 35 days old or older are placed in a short glass tube. The ends of the tube are covered with cheesecloth and it is placed in a automatic tissue processor for about 48 hours. The smaller specimens remain immersed in a finger bowl containing the appropriate fixative which is changed periodically within a similar 48-hour period.

Fixed measurements are made just before the embryos are transferred to 50% or 70% alcohol. Photographs are taken in 70% or 80% alcohol. The embryos are dehydrated by running them through a series of alcohol baths of increasing concentration. Clearing is accomplished in Oil of Wintergreen, Cedarwood Oil, xylene, or more recently in benzene. Specimens cleared in benzene tend to section with less fragmentation. Paraplast, with a melting point of 56°C, is used as the embedding agent. In some cases paraffin, with a melting point of 52°C, is used for embedding the bilaminar and trilaminar stages.

No special procedure, such as Heard (1956) described for Heuser and Streeter, is followed to orient the embryos. The paraffin block is trimmed and then mounted on a small wooden block. The plane of section is determined at the time the block is trimmed and placed in the microtome. All sectioning is done on an AO rotary microtome either in a transverse or sagittal plane at 6 to 10μ. All staining of the embryos is done with Harris's hematoxylin and eosin, with a few exceptions in which protargol-S and eosin is used to demonstrate nerve fibers.

IV. Reconstruction

Three different types of reconstruction are used to illustrate internal developmental characteristics: (A) graphic, (B) wax, and (C) plaster reconstructions.

A. *Graphic*

The graphic reconstructions are made according to the method of Gasser (1967a, b). The serial sections of the embryo to be graphically reconstructed are projected on plain white paper or graph paper using a trisimplex microprojector at a predetermined magnification. The projected image of previously selected sections is measured with a millimeter ruler and recorded. Appropriate measurements are made on the right half of each selected specimen. The rostrodorsal measurement is made first and is taken in the midline of the sections. All other measurements are taken from a line which touches the dorsal border of the section and is perpendicular to the midline. After the sections are measured a graph is constructed which consists of parallel horizontal lines, 1 to 3 mm apart, with every 5th line heavy. The distance between the lines depends on the thickness of the section, their magnification, and the number of sections measured. Each line is labeled with the appropriate slide, row, and section number. The sections with the largest rostrodorsal diameter are the first plotted on the graph followed by sections cranial and caudal to them. Close attention is given to the location of specific organs to obtain accuracy in placing the measurements above one another on the graph. The measurements are first placed on the graph in the form of dots on the parallel lines. When the dots appear to be in the proper horizontal position they are joined with a solid line. Photographs of the embryo are very helpful in getting the proper conformation. The picture constructed in this manner gives a reasonably accurate lateral view of one side of the embryo. Grooves and contours are measured as to their location, and their depth is estimated from the photograph and presented as dark or light shadows.

After the external form and the larger internal structures are arranged on the graph, Xerox copies of the graph are made so that additional structures can be plotted on these copies. The transversely cut serial sections are examined under a microscope and the location, arrangement, and extent of a given structure are plotted on the graph, section by section.

B. *Wax*

The wax reconstructions result in essentially the same type of map as the graphic reconstruction, with the added feature of having definite external contours. The wax reconstructions are made according to Born's method (1883), with some variations. Selected sections are projected onto a white undercoated glass plate at a magnification between $\times 30$ and $\times 100$ by means of a trisimplex microprojector. The sections selected depend upon their thickness and the magnification used. Five-by-ten-inch sheets of dental tray spacer wax are used which have a uniform thickness of 1.7 to 2.3 mm. Its consistency is ideal for cutting with a thin-bladed steel iridectomy knife or a similarly shaped instrument. Each section is projected directly onto a light yellow wax sheet lying on the glass plate. White powder is used to provide greater clarity in those specimens in which the surface ectoderm is lightly stained. The outline of each preselected section is cut through the wax sheet or traced with a sharp-pointed pencil and then cut. The resulting wax impression represents a known number of specific sections. This is determined by dividing the thickness of the wax by the product of the thickness of the projected section (in mm) and the magnification. The slide, row, and section number is lightly carved onto the upper surface of each wax impression.

The wax impressions are then stacked in order. Between every 5th sheet a plastic tab, containing the slide, row, and section number of the wax impression beneath it, is inserted at the rostral, or ventral, and dorsal ends of the impression in such a way that it protrudes several centimeters from the wax model. The sheets are arranged so that the contours correspond with one another. Photographs of the whole embryo are invaluable in the final arrangement of the impressions. Steel T-pins or ordinary stick pins are used to secure the sheets in place. The wax model with a millimeter ruler is photographed from a centered lateral position and the photographs made at a 1:1 magnification. The photograph is traced onto a thin white vellum paper with particular attention given to the location of the plastic tabs. Corresponding dorsal and rostral or ventral tab points are joined on the tracings with a heavy solid line. Each of the lines is roughly parallel to the others. Four lighter parallel lines are drawn between the darker tab lines. All the lines are approximately an equal distance apart (1.7–2.3 mm). Each wax sheet in the model is represented on the drawing as a line which is given its corresponding slide, row, and section number. The extent and depth of the contours of the wax model can be incorporated into the drawing by shading.

The serial sections are then studied under the microscope. Specific structures are plotted on Xerox copies of the drawing in a manner identical to that used for the graphic reconstructions.

C. *Plaster*

Plaster reconstructions are made of internal structures when interpretation of the serial sections is difficult or when it is necessary to illustrate the three-dimensional configuration of an organ, for example, the pharynx or the cochlea. This method is identical to that used for wax reconstruction with the following modifications. The outline of the structure to be studied is cut from wax, using either the wax impressions of previously cut wax models or wax sheets of a predetermined size and thickness based on the magnification selected. The use of wax impressions from existing wax models offers the advantage of alignment consistent with the external form. The resulting wax sheets, which serve as a mold, are stacked in order and small gauge copper wire is placed at appropriate points in the mold to provide strength. The mold is then placed on a vibrator and orthodontic plaster, mixed with water to a thin consistency, is poured into the mold and allowed to set. The wax is melted from the plaster model by placing the mold in an oven at a temperature of 65–70°C for several days. The plaster model is then photographed and drawn.

V. Photography

With the exception of the photomicrographs, all photography, layout, illustration, and artwork is done by the Department of Biomedical Communications at the Southwest Foundation for Research and Education. Photomicrographs are made using a Zeiss Ultraphot II microscope-camera unit. The embryos are photographed using one of three photographic units based on the size of the embryo. Presomite embryos 0.5 to 1 mm are photographed on 4×5 inch film using a Leitz bellows unit equipped with either a 16 or 32 mm microtessar lens. Embryos 1 to 4 mm are photographed using a Nikon F camera equipped with a bellowscope and using a 55 mm F 3.5 Micro-Nikkor lens. Photographs of embryos longer than 4 mm are made using the Nikon F camera equipped with the Micro-Nikkor lens which may be extended with an M-ring. One or two spotlights are used to provide the necessary illumination.

The embryos are prepared for photography as follows: The presomite embryos are mounted in 70% alcohol on a large depression slide which may be temporarily sealed with a coverslip. The embryo is then photographed using the bellows described above. The proper orientation for photography is accomplished by trimming the placenta so it can be used to support the delicate embryo and still get the required views. This is done by cutting a block of the placenta around the embryo on a plane that best illustrates the embryo. The placental block is rotated from one surface to another as it is photographed without damage to the embryo. It is often difficult to expose the embryo on all sides because of its orientation to the placenta; consequently the side apposed to the placenta may not be photographed. The placental block also serves to support the embryo during processing.

The early somite stage embryos are left attached to the placenta in a manner similar to the presomite stages. Embryos 28 days old or older are removed from the placenta by cutting the umbilical cord. They are photographed in a small rectangular tray, made of black plastic, containing 70% alcohol.

References

Born, G. 1883. Die Plattenmodellirmethode. *Arch. Mikr. Anat.* 22:584–89.

Claborn, L. C.; A. G. Hendrickx; and F. H. Kriewaldt 1967. A proved technique for the delivery of early-stage baboon embryos. In *The Baboon in Medical Research,* ed. H. Vagtborg, 2:825–34. Austin: University of Texas Press.

Gasser, R. F. 1967a. The development of the facial nerve in man. *Ann. Otol. Rhinol. and Laryngol.* 76:1–37.

——— 1967b. The development of the facial muscles in man. *Amer. J. Anat.* 120:357–76.

Hartman, C. G. 1944. Regeneration of the monkey uterus after surgical removal of the endometrium and accidental endometriosis. *West. J. Surg. Obstet. Gynecol.* 52:87–102.

Heard, O. O. 1956. Methods used by C. H. Heuser in preparing and sectioning early embryos. *Contrib. Embryol., Carneg. Inst.* 36:1–18.

Marston, J. H.; R. Yanagimachi; M. C. Chang; and D. M. Hunt 1964. The morphology of the first cleavage division in the mouse, Mongolian gerbil, golden hamster, and rabbit. *Anat. Rec.* 148:417 (abstract).

3

Description of Stages I, II, and III

Duane C. Kraemer/Andrew G. Hendrickx

I. Age and Size

STAGE I—One-cell embryo, fertilized ovum, zygote; EFA, 1 day; diameter, 140–180μ; zona pellucida thickness, 12–26μ.

STAGE II—Segmenting embryo; EFA, 2–5 days; diameter 150–200μ; zona pellucida thickness, 14–26μ.

STAGE III—Free blastocyst; EFA, 5–8 days; diameter, 180–200μ; zona pellucida thickness, 9–10μ or absent.

Data on the age, size, and cell number of embryos of Stages I, II, and III are given in table 3.1. There is a close correlation between age and cell number, with cell number increasing with age. The only exceptions to this are embryos A67-210 (II) and A67-110 (III). The discrepancy in age is not greater than one day. It is significant to note that in both of these cases the mating was continuous, thus increasing the chances of variability. The diameter of the preimplantation embryo varies from 140 to 200μ. There is no consistent relationship between age and size for these stages of development although there is a tendency toward expansion of the embryo in Stages II and III. All of the 9 single matings occurred on or within 1 day of the optimal mating time (the 3d day preceding deturgescence) (see chap. 1). This tends to justify the placing of considerable emphasis on the optimal mating time when estimating the fertilization age of embryos from continuous matings as discussed in chapter 2.

II. External Characteristics

The preimplantation period of development is divided into 3 stages on the basis of age and morphological characteristics. These stages begin with the entry of the fertilizing spermatozoan into the vitellus of the ovum and end immediately preceding implantation.

In Stage I the one-cell embryo is surrounded by the zona pellucida and is located in the uterine tube. Corona radiata cells and spermatozoa may be present on the zona pellucida which is often granular in appearance (fig. 3.1B). The segmenting embryo is surrounded by the zona pellucida which is practically devoid of corona radiata cells. The spermatozoa remain visible on the zona pellucida (fig. 3.2). Stage II embryos may be located in either the uterine tube or uterus. By Stage III the embryo lies free in the uterine cavity. The zona pellucida may be

45

either present or absent. If present it is thinner, apparently due to the expansion of the embryo. Late in Stage III the zona pellucida is shed, and the trophoblast becomes the external structure (fig. 3.5).

III. Internal Characteristics

Stained, serial sections of the embryos were examined microscopically. The structures discussed below were examined for their level of development. Each specimen

TABLE 3.1

EMBRYOS OF STAGES I, II, AND III

	INSEM. AGE (Days)	EFA (Days)	MIN. AGE (Days)	DIAMETER (μ)*	ZONA PELL. THICKNESS (μ)*	CHARACTERISTICS†	MATING	
							Single	Cont.
Stage I								
A68-23	1	1	1	155	12	pronuclear	×	
A67-5	1	1	1	178	26	syngamy	×	
Stage II								
A67-16	2	2	2	140	14	4 cells	×	
A66-147	2	2	1	. . .	. . .	6 cells	×	
A66-190	3	3	2	145	16	7 cells	×	
A68-238	6	4	3	158	14	15 cells		×
A67-77	5	5	4	202	26	16 cells	×	
A67-235	5	5	4	196	17	63 cells	×	
A67-210	9	6	5	162	15	64 cells		×
Stage III								
A67-110	10	5	4	195	9	132 cells 89 TB 43 ICM		×
A68-150	6	6	5	183	10	178 cells 130 TB 48 ICM	×	
A66-178	7	7	6	. . .	. . .	. . .	×	
A67-206	10	8	7	200	9	not sectioned		×
A67-222	9	7	6	152	absent	243 cells 199 TB 44 ICM		×

* Measurements were made from photographic negatives.
† TB, trophoblast; ICM, inner cell mass.

was placed subsequently in a particular stage based on the combined level of development of these structures.

A. *Stage I* (Fig. 3.1)

Stage I begins with the entry of the fertilizing spermatozoan into the vitellus of the ovum and continues until the first segmentation division is completed. Two embryos were recovered at this stage. One is in the *pronuclear phase* in which both the male and female pronuclei are recognizable (fig. 3.1*A*). The other is in *syngamy,* the phase in which the 2 pronuclei fuse to form 1 nucleus (fig. 3.1*B*). The diameters of the pronuclei are 15μ and 17μ. Both embryos exhibit 2 polar bodies within the perivitelline space and 1 polar body of the embryo recovered at syngamy appears to be segmenting. Spermatozoa were recovered in the washings of both embryos, but only the embryo at syngamy has spermatozoa attached to the zona pellucida (fig. 3.1*B*).

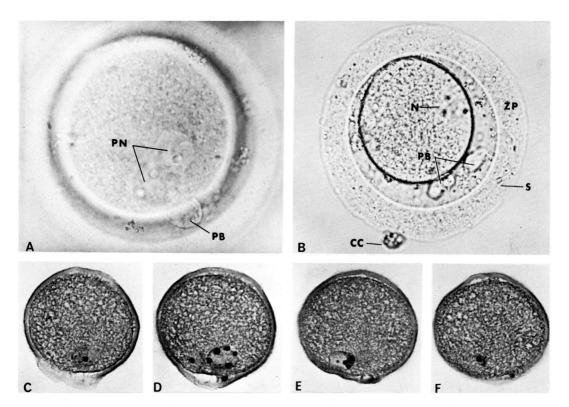

Fig. 3.1. Photomicrographs of Stage I embryos recovered from the uterine tube at day 1 EFA. *A*, whole mounts of pronuclear embryo showing 2 pronuclei (*PN*) and 1 of the 2 polar bodies (*PB*). (×465) *B*, whole mount of embryo at syngamy showing the nucleus (*N*), 2 polar bodies (*PB*), spermatozoa (*S*), granular zona pellucida (*ZP*), and a corona cell (*CC*). (×325) *C–F*, representative sections through the pronuclei and polar bodies of the embryo shown in *A* above. (×360) The erosion of the zona pellucida and shrinking of the cytoplasm explain the differences in size of the specimen in *A* and *C–F*.

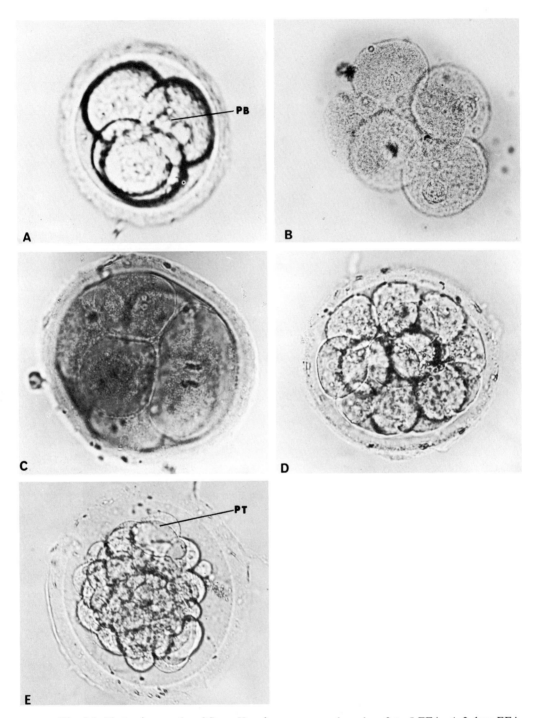

Fig. 3.2. Photomicrographs of Stage II embryos recovered on days 2 to 5 EFA. *A*, 2 days EFA, with 4 blastomeres and 2 polar bodies (*PB*). (×340) *B*, 6-cell embryo recovered from the uterine tube at 2 days EFA. The zona pellucida was dissolved by acetic acid-alcohol fixation. (×405) *C*, embryo recovered from the uterine tube at 3 days EFA, with 7 blastomeres, 1 in the anaphase stage of mitosis. (×405) *D*, a 15-cell embryo recovered from the uterine cavity at 4 days EFA. (×355) *E*, a 63-cell embryo recovered from the uterine cavity at 5 days EFA. (×290) Note the large peripherally located primitive trophoblastic cells (*PT*).

B. *Stage II* (Figs. 3.2, 3.3)

Stage II begins with the completion of the first segmentation division and ends immediately preceding the formation of the segmentation cavity. This stage includes 7 embryos with estimated fertilization ages of 2–6 days. Development during days 2 (4 and 6 cells) and 3 (7 cells) occurs in the uterine tube. The first recovery of embryos from the uterine cavity was on day 4 (15 cells). Cell differentiation was first observed in the 63-cell embryo (day 5, from uterine

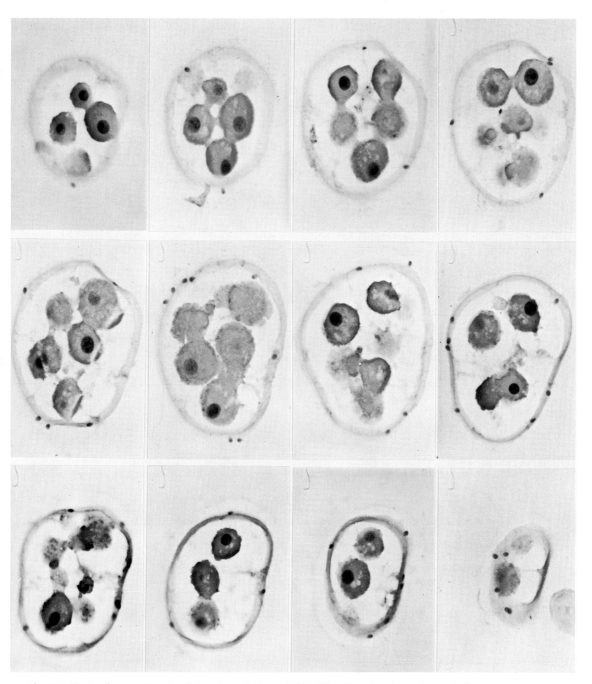

Fig. 3.3. Photomicrographs of serial sections 2 through 13 of the 15-cell embryo shown in figure 3. 2D. (×360)

cavity) in which large primitive trophoblastic cells appear at the periphery of the morula (fig. 3.2E).

C. *Stage III* (Figs. 3.4, 3.5)

Stage III begins with the formation of the segmentation cavity and ends with the first signs of implantation. The 5 embryos range in their estimated fertilization ages from 5 to 8 days. The length of this stage is approximately 3 days longer in the baboon than has been estimated for man (Hertig, Rock, and Adams 1956). During these 3 days there is little, if any, increase in the number of cells in the inner cell mass, whereas the number of trophoblastic cells is more than doubled. Shedding of the zona pellucida from the blastocyst occurs on days 7 or 8 in the baboon in contrast to estimated day 5 in man.

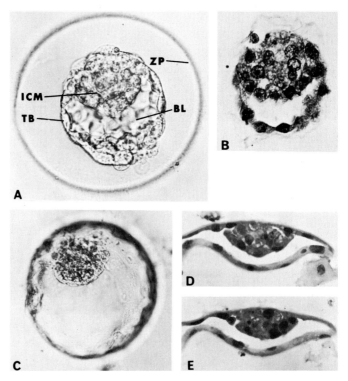

Fig. 3.4. Photomicrographs of Stage III embryos before loss of the zona pellucida. *A*, free blastocyst recovered from the uterine cavity at 5 days EFA. (×265) Note the blastocoel (*BL*), inner cell mass (*ICM*), trophoblast (*TB*), and the zona pellucida (*ZP*). The embryo is contracted and pulled away from the inner surface of the zona pellucida apparently due to hypertonicity of the lactated Ringer's collection fluid. *B*, section from the same embryo through the inner cell mass. (×360) *C*, free blastocyst recovered from the uterine cavity at 6 days EFA. Since this embryo was collected in 0.9% saline, the trophoblast is not contracted and lies adjacent to the inner side of the zona pellucida. (×220) *D*, *E*, sections of the inner cell mass of the same embryo as shown in *C*. (×325) The trophoblast has collapsed during processing so that its position ventral to the inner cell mass is artifactual.

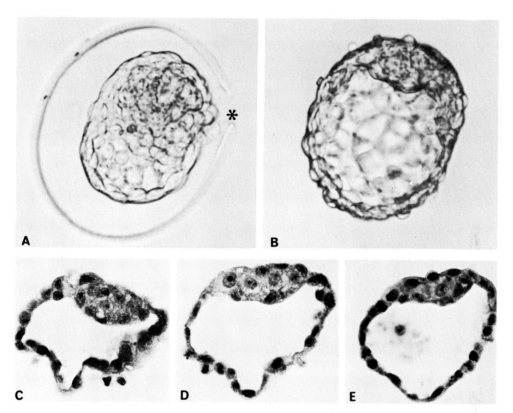

Fig. 3.5. Photomicrographs of Stage III embryos during and after shedding of the zona pellucida. *A*, a free blastocyst recovered from the uterine cavity at 8 days EFA. (×290) The zona pellucida is in the process of being shed. Asterisk(*) indicates the opening in the zona pellucida. *B*, a free blastocyst recovered from the uterine cavity at 7 days EFA. (×295) The zona pellucida has been shed. *C–E*, sections through the inner cell mass of the same embryo as shown in *B*. (×360)

References

Hamilton, W. J. 1949. Early stages of human development. *Ann. Roy. Coll. Surg. Engl.* 4:281–94.

Hendrickx, A. G., and D. C. Kraemer 1968. Preimplantation stages of baboon embryos (*Papio* sp.). *Anat. Rec.* 162:111–20.

Hertig, A. T.; J. Rock; E. C. Adams; and W. J. Mulligan 1954. On the preimplantation stages of the human ovum: a description of 4 normal and 4 abnormal specimens ranging from the 2d to the 5th day of development. *Contrib. Embryol., Carneg. Inst.* 35:201–20.

Hertig, A. T.; J. Rock; and E. C. Adams 1956. A description of 34 human ova within the first 17 days of development. *Amer. J. Anat.* 98:435–93.

Heuser, C. H., and G. L. Streeter 1941. Development of the macaque embryo. *Contrib. Embryol., Carneg. Inst.* 29:17–51.

4

Description of Stages IV, V, VI, VII, and VIII

Andrew G. Hendrickx/Marshall L. Houston

I. Age and Size

STAGE IV—Implanting blastocyst, inner cell mass; EFA, 9 ± 1 days; length, 0.04 mm.

STAGE V—Blastocyst implanted but still avillous, bilaminar embryonic disc; EFA, 10 ± 1 days; length, 0.08 mm.

STAGE VI—Primitive chorionic villi, bilaminar embryonic disc; EFA, 11–15 days; length, 0.12–0.20 mm.

STAGE VII—Branching chorionic villi, trilaminar embryonic disc; EFA, 16–18 days; length, 0.20–0.35 mm.

STAGE VIII—Angioblasts in chorionic villi, primitive streak; EFA, 19–21 days; length, 0.60–0.90 mm.

Data on the age and size of the embryos of Stages IV through VIII are given in table 4.1. The embryos are listed in the order of their morphological development. There is a close correlation between the estimated fertilization age and the morphological development of embryos belonging to Stages IV, V, and VI. In Stages VII and VIII there is little correlation, although there is a tendency for an advance in development with age. Age and size are also relatively well correlated; however, there are some inconsistencies. This is explained in part by the use of different fixatives, which increases the chances of variability. Specimen A66-126 (VII) is small and specimens A65-193 (VI), A65-135 (VII), and A65-124 (VIII) are large. Two of these embryos, A65-193 and A65-135, are among the more advanced embryos of Stages VI and VII, respectively, and A65-124 is one of the less advanced of Stage VIII.

The range in embryo length increases progressively in Stages VI (0.12–0.34 mm), VII (0.16–0.50 mm), and VIII (0.57–1.0 mm). The increase in length from stage to stage is quite consistent even for Stage VII where embryos tend to be slightly smaller. The spurt of growth in Stage VIII may be accounted for by the formation and proliferation of the primitive streak. In addition, after Stage VII there appears to be a balance between the growth of the embryo and the placenta.

Of the 13 single matings, 9 occurred on or within 1 day of the optimal day for mating (the 3d day preceding deturgescence). Four occurred earlier; 1 on day 5 (A68-264, V), 1 on day 6 (A65-152, VIII), and 2 (A65-165, VII, and A65-164, VIII) on day 7 preceding deturgescence. The early matings along with the different fixatives that were used increase the chance for variability. The estimated fertilization age of the embryos from the early matings corresponds more closely with the developmental characteristics for the respective stages than does the insemination age.

II. External Characteristics

Data on the sites of implantation and chorionic diameters for the embryos of Stages IV through VIII are given in table 4.2. For purposes of locating and recording the implantation sites, the wall of the uterine cavity was divided into *dorsal* and *ventral* surfaces. Each surface was subdivided into *cranial* (upper) and

TABLE 4.1

AGE, SIZE, AND MATING FOR EMBRYOS OF STAGES IV–VIII

	INSEM. AGE (Days)	EFA (Days)	MIN. AGE (Days)	GREATEST LENGTH, WIDTH, THICKNESS (mm)*	MATING	
					Single	Cont.
Stage IV						
A68-154	9	9	8	.04×.04×.03	×	
Stage V						
A68-264	11	9	9	.08×.07×.02	×	
Stage VI						
A67-163	11	11	10	.12×.10×.03	×	
A67-156	15	12	11	.17×.13×.04		×
A67-99	13	13	12	.14×.12×.03	×	
A65-193	15	15	14	.34×.31×.06		×
A65-151	16	15	14	.23×.20×.07		×
Stage VII						
A65-187	16	16	16	.19×.19×.06	×	
A66-126	19	18	17	.16×.14×.07	×	
A68-18	. . .	17	16	.21×.21×.04		×
A65-135	. . .	16	15	.50×.40×.06		×
A65-165	20	18	18	.30×. . .×.06	×	
A65-168	. . .	18	17	.35×.21×.04		×
Stage VIII						
A65-138	21	20	19	.57×.45×.06	×	
A65-183	. . .	19	18	. . . .		×
A65-136	. . .	21	20	.75×.74×.06		×
A65-124	21	21	20	1.0 ×.65×.08	×	
A65-141	26	20	19	.78×.62×.09		×
A65-152	22	20	20	. . .×.78×.14	×	
A65-164	23	21	21	.90×.57×.07	×	
A65-173	21	20	19	.89×.75×.09	×	
A65-185	26	21	20	. . . .		×
A65-202	24	23	22	. . . .	×	

* Measurements were made postfixation from sections.

caudal (lower) halves with *left, central,* or *right* positions. Twelve of the 23 embryos were implanted on the dorsal surface and 10 were implanted on the ventral surface (the implantation site for 1 embryo was not recorded). Implantation occurred almost twice as often in the cranial half (14) of the uterine wall as in the caudal half (8). Eighteen of the 22 implantations had a central position, 3 were on the left side, and 1 was on the right side.

Based on the identification of the corpus luteum at laparotomy, ovulation was equal in the two ovaries. A crossover occurred in only one case, A65-193 (VI), where the right ovary ovulated but the embryo implanted on the left side. However, there was a groove on the surface of the endometrium which resulted from a previous embryotomy and may have altered the course of this implanting embryo. The diameter of the chorion increases rather consistently in each stage.

At implantation (Stage IV) the blastocysts are oriented with the inner cell mass adjacent to the uterine surface (fig. 4.1). The abembryonic wall of the blastocyst is collapsed in the younger specimens and it is difficult to determine if this collapse is a natural occurrence as suggested by Wislocki and Streeter (1938) or is an artifact. The blastocysts of Stages IV and V are detectable as small, whitish masses on the endometrial surface. In Stage VI the embryonic disc is round to oval shaped and remains in close apposition to the chorionic plate (fig. 4.3). Its dorsal surface lies parallel to the surface of the placenta, and the body stalk differentiates by a condensation of the mesenchyme. In Stage VII the embryonic disc elongates, takes on a convex, oval shape, and protrudes slightly into the

TABLE 4.2

CHORIONIC DIAMETER, IMPLANTATION SITE, AND LOCATION
OF CORPUS LUTEUM FOR EMBRYOS OF STAGES IV–VIII

	Chorionic Diameter (mm)*	Implantation Site	Location of Corpus Luteum
Stage IV			
A68-154	0.18	dorsal, upper, central	
Stage V			
A68-264	0.37	. . . .	right
Stage VI			
A67-163	0.67	dorsal, upper, left	left
A67-156	0.78	dorsal, upper at utero-tubal junction, right	right
A67-99	1.27	ventral, lower, central	left
A65-193	2.47	ventral, lower, left	right
A65-151	1.61	dorsal, upper, central	left
Stage VII			
A65-187	2.00	ventral, upper, central	left
A66-126	2.18	ventral, upper, central	right
A68-18	2.08	dorsal, upper, central	left
A65-135	5.72	dorsal, upper, central	left
A65-165	7.3	ventral, lower, central	left
A65-168	3.9	dorsal, upper, central	right
Stage VIII			
A65-138	9.5	dorsal, upper, central	right
A65-183	14.0	ventral, upper, right	right
A65-136	9.6	dorsal, upper, central	right
A65-124	13.0	ventral, lower, central	right
A65-141	14.0	ventral, lower, central	left
A65-152	11.0	dorsal, upper, central	right
A65-164	13.0	dorsal, lower, central	right
A65-173	15.0	ventral, lower, central	left
A65-185	. . .	ventral, upper, central	left
A65-202	4.5	dorsal, lower, central	left

* Measurements on the specimens in Stages IV through VII were made postfixation from sections. Stage VIII specimens were measured grossly after fixation.

amniotic cavity (figs. 4.4, 4.5, 4.6). By Stage VIII the disc becomes more elongated and its dorsal surface is convex, having curvatures in the transverse and longitudinal planes (fig. 4.8). The embryos are attached to the placenta by a broad and sometimes extensive body stalk.

The implantation site of the Stage IV blastocyst appears as a small, transparent dome which is raised only slightly above the uterine surface. Small quantities of extravasated blood are usually present around its outer border. Little change occurs in the implantation site in Stage V, but by Stage VI it is expanded in diameter and the chorionic sac is elevated above the uterine epithelium (see chap. 9, fig. 9.1). In the fresh state the implantation site is a pinkish red color, probably produced by the blood in the lacunar spaces as well as the hemorrhaging of vessels in the placenta. The surrounding endometrium is a contrasting pink color. In the younger embryos of Stage VII the placenta remains elevated above the uterine epithelium. In the older specimens it begins to overgrow the epithelium laterally and is depressed centrally. The central portion of the placenta is very depressed in Stage VIII.

III. Internal Characteristics

Stained, serial sections of the embryos were examined microscopically. The structures discussed below were examined for their level of development. Each specimen was placed subsequently in a particular stage based on the combined level of development of these structures.

A. *Stage IV* (Fig. 4.1)

1. *Inner cell mass.* The embryonic disc is not yet evident in the inner cell mass which is attached to the wall of the blastocyst where the trophoblast is attached to the endometrium (fig. 4.1*A–C*). The cells of the inner cell mass show no organization and are actively proliferating. The larger portion of these cells appears to be epiblastic but occasional endoblastic cells are apparent on that aspect of the inner cell mass that borders the blastocele. The endoblastic cells have more oval-shaped nuclei but are not arranged into a distinct layer.

2. *Extraembryonic membranes.* The blastocele is larger than in Stage III but, as yet, the amniotic cavity, vitelline cavity, and allantois are not evident.

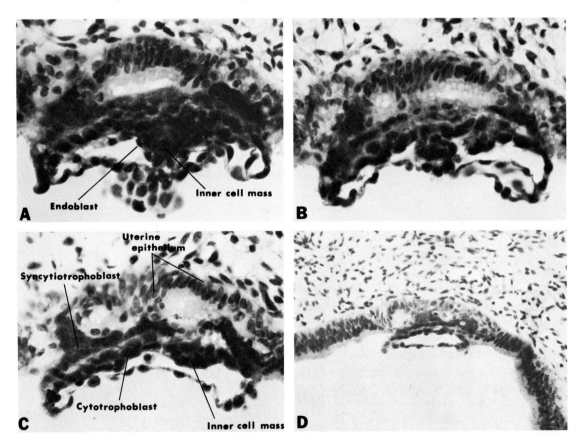

Fig. 4.1. Photomicrographs of the implanting blastocyst of Stage IV. The abembryonic portion of the trophoblast is collapsed against the inner cell mass. *A–C*, sections through the points of coalescence between the blastocyst and the uterine epithelium. (×500) *D*, section through the uterine stroma and implantation site. (×200)

3. *Trophoblast and placenta.* The trophoblast over the free, abembryonic pole of the blastocyst is a single, cellular layer, but in the area of contact with the uterine epithelium it is distinctly differentiated into cytotrophoblast and syncytiotrophoblast. The inner, cytotrophoblastic layer, which is contiguous with the inner cell mass, is continuous with the abembryonic trophoblast and is only one or two cells thick. The syncytiotrophoblast is located externally on the embryonic pole of the blastocyst and is in contact with the uterine epithelium (see fig. 9.2). In the areas of contact the syncytiotrophoblast is thicker. The uterine epithelial cells are unorganized and lose their normal columnar appearance. As much as one-half of the surface portion of these cells has disappeared.

B. *Stage V*

1. *Bilaminar embryonic disc I.* Only one specimen was collected that has features characteristic of this stage. It is in poor condition and is not illustrated. The embryonic disc is evident for the first time and is becoming bilaminar. Its orientation is unusual in that it lies at right angles rather than parallel to the endometrial surface. Epiblastic cells are becoming columnar and are forming a plate that is slightly convex toward the amniotic cavity. The endoblast is poorly defined on the ventral aspect of the plate.

2. *Extraembryonic membranes.* The amniotic membrane is composed of a single layer of cells and together with the embryonic disc encloses the small, slit-like amniotic cavity. As yet, there is no indication of the vitelline sac and allantois.

3. *Trophoblast and placenta.* The trophoblast forms a thick plate in the area of attachment to the uterine wall where it is composed primarily of syncytiotrophoblast (see fig. 9.3). The cytotrophoblast is no more than two cell layers thick over the embryonic surface and a single layer on the abembryonic wall of the blastocyst. Small clefts within the syncytiotrophoblast are precursors of the trophoblastic lacunae. The blastocyst has completely penetrated the uterine epithelum and lies partly in the stroma of the endometrium. The trophoblastic penetration attains its greatest depth in the necks of uterine glands that are covered by the implantation site. The uterine stroma is edematous, and superficial capillaries deep to the implantation site increase in number.

C. *Stage VI* (Figs. 4.2, 4.3)

1. *Bilaminar embryonic disc II.* The well-defined, bilaminar, embryonic disc is composed of an epiblastic plate that borders the amniotic cavity and an endodermal layer that borders the vitelline cavity (fig. 4.2*A, B*). The organization of the cells of the epiblast into pseudostratified, columnar cells is more evident in the older specimens (fig. 4.2). Two to three layers of nuclei are evident in the plate and the height of the cells diminishes near the periphery of the plate. The change in cell type, from columnar to squamous, at the junction of the epiblastic plate and the amnion is rather abrupt in younger embryos, whereas in the older specimens this transition is more gradual.

The extent of the endoderm varies from several scattered cells to a definite layer. Endoderm lies adjacent to the epiblast in some areas, but a cleft that is thought to be an artifact sometimes separates them (fig. 4.2). Cells are sometimes observed in the vicinity of the cleft and appear to be migrating from the epiblastic plate into the endoderm (fig. 4.2*C, D*). The endodermal layer is not as thick as the epiblastic plate and does not always extend to the periphery of the embryonic disc. It is continuous with the cell layer forming the primitive vitelline cavity. The endoderm will form the gut primordium and its cells are somewhat smaller, less well organized, and more loosely arranged than those in the epiblastic plate. Cell boundaries are difficult to see. The nuclei are irregularly shaped and contain nucleoli in the older specimens. There is an occasional suggestion of alignment of the cells into the platelike arrangement seen in later stages. In one embryo (A65-156) a region of the endoderm is pseudostratified and probably represents the first sign of the prochordal plate that establishes the craniocaudal axis and divides the disc into right and left sides.

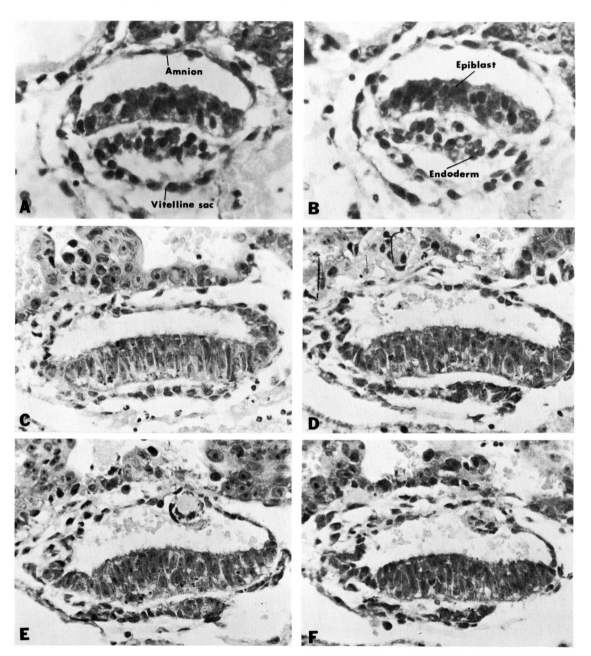

Fig. 4.2. Photomicrographs of a bilaminar blastocyst of Stage VI. *A,B,* sections through an 11-day embryo. (×415) *C–F,* sections through a 15-day embryo. (×240) The vitelline sac is particularly well developed in the younger embryo but appears to be lagging in the older specimen. The cells of the epiblastic plate of the younger embryo are loosely arranged while those of the older one are more condensed. The endodermal plate of the younger embryo, although loosely arranged, is more abundant than that in the older specimen. Cellular concretions and erythrocytes appear in the amniotic cavity at *D–F.*

2. *Extraembryonic membranes.* The vitelline cavity on the underside of the embryonic disc is smaller than the amniotic cavity and varies from a small slit to a dome-shaped space (fig. 4.2). It is enclosed by a squamous cell layer with nuclei that are darkly staining.

The amnion is a double layer of cells which blend with the cytotrophoblast. Little if any mesoblast intervenes between the two layers. The inner, trophoblastic, ectodermal layer consists of squamous cells whose nuclei are darkly staining and rather tightly arranged so that only small amounts of cytoplasm intervene. The outer, mesothelial-like layer is very loosely arranged and contains scattered, darkly staining nuclei. The cells of the amnion closely resemble those of the cytotrophoblast. The amniotic cavity is a dome-shaped space between the embryonic disc and the covering trophoblast and is larger than the vitelline cavity.

A short, ill-defined, connecting stalk composed of irregular clusters of cells can be identified. The magma reticulare is identifiable in all the specimens as cells that extend across the unattached pole of the embryo and blend with the inner layer of the trophoblast (fig. 4.2D, F).

3. *Trophoblast and placenta.* The chorionic vesicle consists of a villous portion where cords of trophoblast project from the surface applied to the uterine wall, forming the placenta, and a nonvillous or membranous portion that comprises the unattached area of the vesicle. The membranous chorion is a delicate, double-layered membrane consisting of two closely apposed layers; an outer trophoblast and an inner mesothelial layer. It is transparent although the embryo is not identifiable through it.

By the beginning of Stage VI, the placenta is predominantly cytotrophoblast. The surface of the cone-shaped mass is level with the internal surface of the uterine mucosa (fig. 4.3A, B). The trophoblast lacunae are large, filled with maternal blood, and have begun to coalesce into an interconnecting space. Short, straight villi, containing numerous mesoblasts, extend from the placental surface, particularly in the area deep to and immediately surrounding the embryo. Heavy columns of cytotrophoblastic cells continue from the tips of the open villi to the base of the placenta.

By the end of this stage the placenta has more than doubled in thickness, pushing out as a mushroom-shaped elevation over the surface of the uterine epithelium (fig. 4.3C, D). The diameter has increased more than 3-fold. The villi progressively elongate, extending half the total depth of the placenta. Some demonstrate rudimentary primary branches. The intervillous space becomes increasingly complete by further coalescence of trophoblastic lacunae. Lateral expansions of cytotrophoblastic cells from the distal tips of the cell columns unite with similar expansions from other cell columns to form a solid trophoblastic shell against the maternal stroma. No decidualization of maternal tissue is evident.

D. *Stage VII* (Figs. 4.4 to 4.7)

1. *Embryonic shield I.* The configuration of the embryonic disc varies from slightly convex to mildly S-shaped. The pseudostratified, epiblastic plate is thicker than in Stage VI but otherwise remains unchanged. Since a third layer of cells is forming between the epiblastic plate and the endoderm, the epiblastic plate is now called ectoderm or the ectodermal plate.

The endodermal layer of the youngest embryo (A65-187) is organized into a plate composed of one to two layers of cells (fig. 4.4). In the older embryos the endodermal plate has expanded and extends almost to the periphery of the ectodermal plate. The prochordal plate begins as a localized thickening of the endoderm and denotes the cranial pole of the embryo (fig. 4.4B). Polarity of the embryo is established with the formation of the prochordal plate at the cranial end and the location of the body stalk at the caudal end. Other than the prochordal plate area, embryonic endoderm is reduced to a loosely arranged, cuboidlike layer of cells that is continuous peripherally with the single-layered vitelline sac.

Stage VII marks the first appearance of the primitive streak (fig. 4.5*D–F*). In its initial development the primitive streak is an area of proliferation along the craniocaudal axis in the more caudal regions of the ectodermal plate. Characteristic of the primitive streak is the disorganization of cells, but this is not apparent

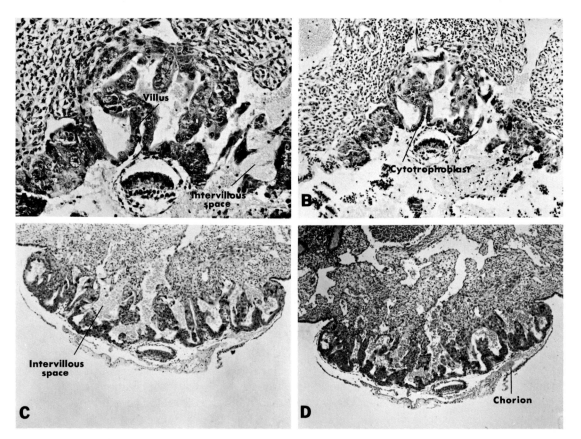

Fig. 4.3. Photomicrographs of the placentae of embryos belonging to Stage VI. *A*, section of an 11-day embryo. (×140) *B*, section of the same embryo. (×85) *C,D*, sections of a 15-day embryo. (×40) The chorionic villi are short, straight, open extensions into the cytotrophoblastic cell columns. The embryonic surface of the 11-day specimen is disrupted. The point of disruption is in direct line with the opening of a maternal blood vessel into the intervillous space and may have been caused by the increased pressure exerted during the surgical removal of the embryo.

throughout the primitive streak areas. Intraembryonic mesodermal cells are sparsely scattered between ectoderm and endoderm in the younger specimens forming a trilaminar embryonic disc. It becomes organized into a distinct layer in the older specimens (fig. 4.5*B–G*). There is no indication of a primitive pit or groove but there is an elevation of the ectoderm in the oldest embryo which gives the location of the primitive streak on the surface of the disc.

2. *Extraembryonic membranes.* The vitelline sac approaches the size of the amniotic cavity (figs. 4.4, 4.6). The wall of the vitelline sac is made up of a single layer of very thin squamous cells and attaches to the periphery of the endodermal plate. The double-layer arrangement of the amnion is well defined. The allantois has not yet formed.

3. *Trophoblast and placenta.* The placenta remains elevated above the level of the free surface of the uterine epithelium, protruding into the uterine cavity. A thin membranous chorion extends from its edges into the uterine cavity (fig. 4.7*A*).

In younger specimens of this stage, chorionic mesoblasts have increased in number both within the open villi and on the placental surface where they become layered. Cytotrophoblastic cells of the placental surface are oriented into a single cuboidal layer which extends distally around the open portions of the trophoblastic villi (fig. 4.7*C*). A few of the villi, particularly near the embryo, show very early stages of branching. Large cell columns are still distinct from the tips of the open villi to the placental base. The cytotrophoblastic shell is complete, and a distinct junctional zone of mixed cytotrophoblasts and necrotic maternal cells has

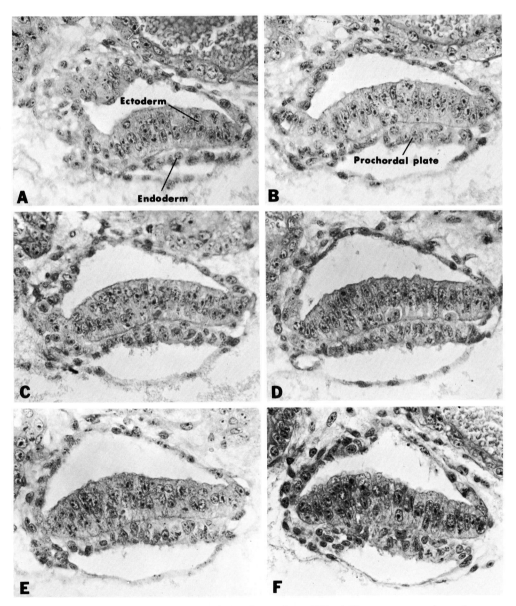

Fig. 4.4. Photomicrographs of an embryo belonging to Stage VII. *A–F,* transverse sections of a 16-day embryo. (×285) Mitotic figures are present, the ectodermal plate is larger, and the columnar shape of the ectodermal cells is well defined. Apparently, some cells are about to migrate from the ectodermal plate to form the presumptive mesodermal layer in *B–D.*

formed against its distal surface. The syncytiotrophoblast is confined to a continuous layer lining the inner surface of the intervillous space.

In the older stages the placenta has begun to overgrow the uterine epithelium laterally and has become slightly depressed centrally. Definite lines of vasoformative mesoblasts (angioblasts) are present on the placental surface and in some specimens extend into the villous cores. The trophoblastic villi extend almost the entire depth of the placenta and are distinctly branched. Some of the larger villi near the embryo show multiple primary branches and very early evidences of secondary branching are present in the more distal regions. Both the trophoblastic shell and junctional zone are increased in thickness and decidualization of the uterine stroma has begun near the implantation site.

The mesoblast separating the amnion from the chorionic plate has increased over that of the previous stage, providing a spongy intervening cushion between these structures. A body stalk is not recognized in the youngest embryos, but an accumulation of mesoblastic tissue at the caudal end of the embryo marks the position of the primitive body stalk by 17 days. The mesoblast of the body stalk is slightly more condensed than the mesoblast between the amnion and the chorionic plate.

E. *Stage VIII* (Fig. 4.8)

1. *Embryonic shield II.* The craniocaudal axis is easily determined at this stage. The ectodermal surface of the caudal pole of the disc is smooth and convex and the prochordal plate is easily located in the endoderm. A condensed body stalk, the cloacal membrane, and a small cloaca are present at the caudal pole.

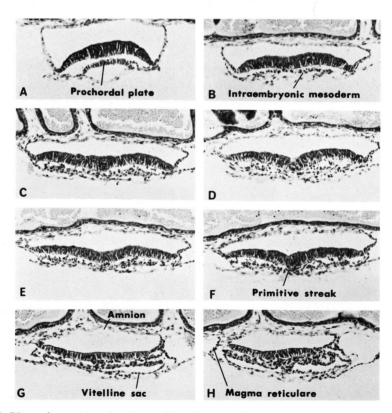

Fig. 4.5. Photomicrographs of a Stage VII embryo. *A–H,* transverse sections of an advanced 16-day embryo. (×100)

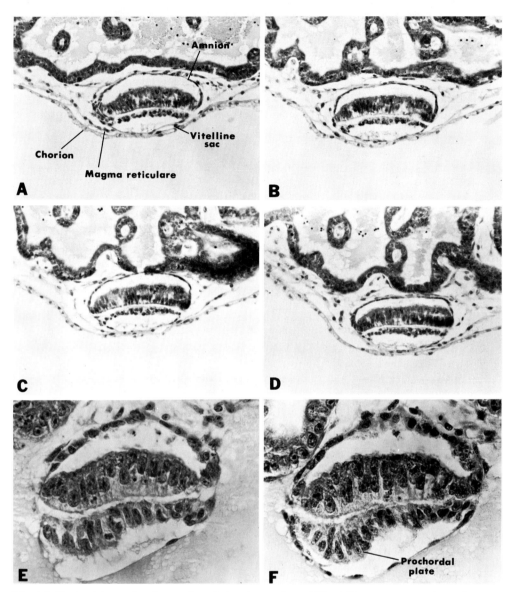

Fig. 4.6. Photomicrographs of an embryo belonging to Stage VII. *A–D*, transverse sections of a 17-day embryo. (×125) *E,F*, longitudinal sections of an 18-day embryo. (×335)

The ectodermal and endodermal layers differ only slightly from Stage VII. The ectodermal plate consists of three to four layers of cells but thins to two and then one layer caudally. Its cells become cuboidal at the periphery of the disc. The smooth convexity of the ectodermal plate is modified through its cranial half by the elevation formed by the notochordal plate and, more caudally, by the primitive streak. The ectodermal epithelium is reduced to two layers above the notochordal plate. It has been modified into a neural groove in only one embryo (A65-173). This same embryo also displays a well formed foregut and is considered to be precocious.

The endodermal layer of the younger embryos is being modified cranially by the enlarging prochordal plate and caudally by the developing cloaca that is a small cul-de-sac. The cells of the prochordal plate are columnar in contrast to the adjacent cuboidal cells. At the periphery of the disc the endoderm becomes a thin layer of squamous cells that are continuous with the lining of the vitelline sac.

The mesodermal layer is abundant and is found throughout the embryo. It is least apparent in the cranial third of the embryo and is most apparent at the level of the primitive streak, especially in the older specimens. Cranial to and at the level of the prochordal and notochordal plates, the mesoderm is in the form of individual cells or small clusters. Adjacent to the cranial part of the primitive

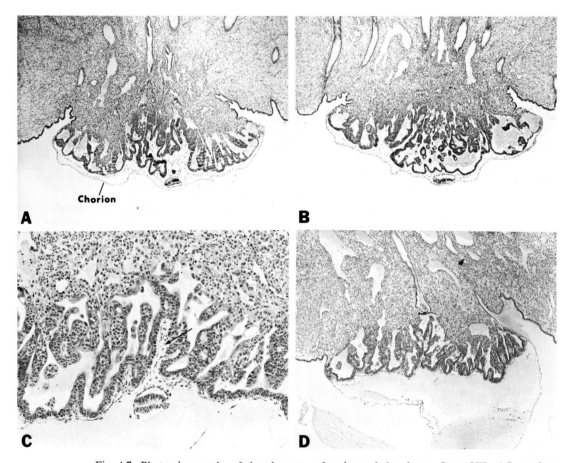

Fig. 4.7. Photomicrographs of the placentae of embryos belonging to Stage VII. *A,B,* sections of the placenta of a 17-day embryo. (×25) *C,* sections of the placenta of an 18-day embryo. (×65) *D,* section of the placenta of the same specimen as *C.* (×25) The placenta is slightly convex in shape and is raised above the surface of the uterine epithelium. Note in *C* the increase in the intervillous space and the distinctly branching villi that contain vasoformative mesoblasts (*arrow*).

streak there is a gradual increase in cells that form the paraxial mesoderm (presumptive somitic mesoderm). Lateral to this paraxial column the mesoderm thins but remains a multilayered plate. Small, isolated separations are visible in the lateral plate mesoderm and will form the intraembryonic coelom (fig. 4.8*J*).

The notochordal plate appears in this stage as a short but moderately thick column of cells extending cranially from the primitive streak. Soon after its appearance it is organized into a rodlike mass (fig. 4.8*H*). This mass is in apposition to the endoderm over its entire length and to the neural plate ectoderm at its cranial end. The notochordal plate extends from the prochordal plate to the primitive streak. The notochordal canal is in the early stages of formation in the oldest embryos.

The primitive streak is a distinct landmark and may occupy as much as three-fifths the length of the embryo (fig. 4.8*I–L*). It is composed of unorganized cells that extend from the notochordal plate to the cloacal membrane. The caudal portion of the streak is smaller and less conspicuous because of the reduced number of cells. It terminates near the cloacal membrane that is formed by the apposition of the ectoderm and endoderm (fig. 4.8*R*). A short, shallow, primitive groove is becoming apparent in the midline on the ectodermal surface of the embryo (fig. 4.8*N*).

2. *Extraembryonic membranes.* Concomitant with the formation of the cloacal membrane the allanto-enteric diverticulum (cloaca) develops in all but the youngest embryos. In the oldest embryos this diverticulum is divided into a hindgut and an allantois. The endodermal layer that forms the inner lining of both these structures is continuous with, but slightly thicker than, the lining of the vitelline sac. The

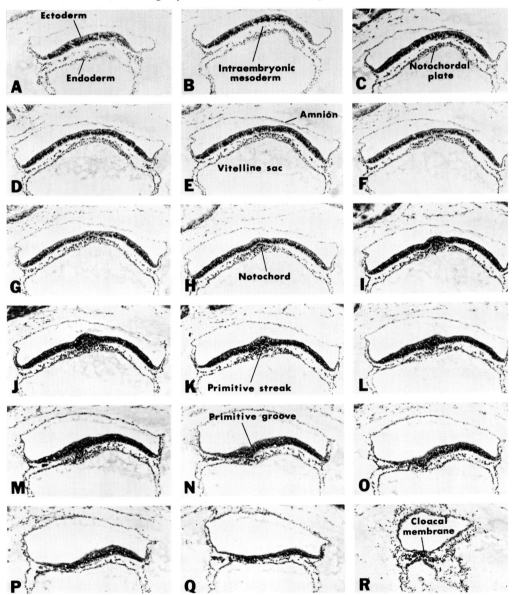

Fig. 4.8. Photomicrographs of a Stage VIII embryo. *A–R,* representative, transverse sections in serial order of a 22-day embryo. (×55)

allantois projects deep into the body stalk which is well formed at this stage. Small diverticula extend from its walls.

The amnion is unchanged from the previous stage. The intermingling of intra- and extraembryonic mesoderm is clearly evident. The amniotic cavity is relatively small and the vitelline cavity is enlarged considerably. The junction of the amnion and the vitelline sac with the embryonic disc is sharply demarcated by a constriction at the edge of the disc. The mesoderm forming the outer layer of the vitelline sac has increased considerably and blood islands are irregularly spaced throughout its walls.

3. *Trophoblast and placenta.* The placenta grows laterally over the surface of the uterine epithelium and begins to be depressed centrally, becoming deeply umbilicated by 25 days (see fig. 9.10*A, B*). The chorionic villi undergo secondary and tertiary branching and reach almost the entire depth of the placenta. By the end of this stage lines of the vasoformative mesoblasts are prominent in the body stalk, placental surface, and villous cores even in the youngest embryos. However, these capillaries contain no blood cells. The cytotrophoblastic shell is several cell layers in thickness, and the junctional zone forms a remarkably smooth line of demarcation between embryonic and maternal tissue (fig. 9.10*E*).

References

Brewer, J. I. 1938. A human embryo in the bilaminar blastodisc stage (the Edward-Jones-Brewer ovum). *Contrib. Embryol., Carneg. Inst.* 27:85–93.

Gilbert, C., and C. H. Heuser 1954. Studies in the development of the baboon (*Papio ursinus*). *Contrib. Embryol., Carneg. Inst.* 35:13–165.

Hamilton, W. J., and J. D. Boyd 1960. Development of the human placenta in the first 3 months of gestation. *J. Anat.* 94:297–328.

Hertig, A. T., and J. Rock 1941. Two human ova of the pre-villous stage, having an ovulation age of about 11 and 12 days respectively. *Contrib. Embryol., Carneg. Inst.* 29:129–56.

———— 1945. Two human ova of the pre-villous stage, having a developmental age of about 7 and 9 days respectively. *Contrib. Embryol., Carneg. Inst.* 31:67–84.

———— 1949. Two human ova of the pre-villous stage, having a developmental age of about 8 and 9 days respectively. *Contrib. Embryol., Carneg. Inst.* 33:171–86.

Hertig, A. T.; J. Rock; E. C. Adams; and W. J. Mulligan 1954. On the pre-implantation stages of the human ovum: a description of 4 normal and 4 abnormal specimens ranging from the 2d to the 5th day of development. *Contrib. Embryol., Carneg. Inst.* 35:199–220.

Hertig, A. T.; J. Rock; and E. C. Adams 1956. A description of 34 human ova within the first 17 days of development. *Amer. J. Anat.* 98:435–93.

Heuser, C. H. 1932. A presomite embryo with a definite chorda canal. *Contrib. Embryol., Carneg. Inst.* 23:253–67.

Heuser, C. H.; J. Rock; and A. T. Hertig 1945. Two human embryos showing early stages of the definitive yolk sac. *Contrib. Embryol., Carneg. Inst.* 31:87–99.

Heuser, C. H., and G. L. Streeter 1941. Development of the macaque embryo. *Contrib. Embryol., Carneg. Inst.* 29:17–55.

Houston, M. L. 1969. The villous period of placentogenesis in the baboon (*Papio* sp.). *Amer. J. Anat.* 126:1–16.

Jones, H. O., and J. I. Brewer 1941. A human embryo in the primitive streak stage (Jones-Brewer ovum I). *Contrib. Embryol., Carneg. Inst.* 29:159–65.

Noback, C. R.; G. H. Paff; and R. J. Poppiti 1968. A bilaminar human embryo. *Acta Anat.* 69:485–96.

Wilson, K. M. 1945. A normal human ovum of 16 days development. The Rochester ovum. *Contrib. Embryol., Carneg. Inst.* 31:103–6.

Wislocki, G. B., and G. L. Streeter 1938. On the placentation of the macaque (*Macaca mulatta*), from the time of implantation until the formation of the definitive placenta. *Contrib. Embryol., Carneg. Inst.* 27:1–66.

5

Description of Stages IX, X, and XI

Andrew G. Hendrickx

I. Age and Size

STAGE IX—0–3 paired somites, formation of neural folds; EFA, 23 ± 1 days; greatest length 1.0–2.0 mm.
STAGE X—4–12 paired somites; EFA, 25 ± 1 days; greatest length, 2.0–3.5 mm.
STAGE XI—13–20 paired somites; EFA, 27 ± 1 days; greatest length, 2.0–4.5 mm.

Data on the age and size of the embryos of Stages IX through XI are given in table 5.1 where they are listed in order of their morphological development. There is a close correlation between the estimated fertilization age and the stage of morphological development in all but 4 cases, A65-150 (IX), A65-172 (IX), A65-176 (X), and A65-179 (X), which were all continuous matings. The estimated fertilization age of A65-150 is equal to that of Stage X embryos but no somites are formed. The estimated fertilization age of A65-172 is comparable to the average age of Stage XIV embryos, which is 6 days more than the average age for Stage IX. On the other hand, cases A65-179 and A65-176 are, respectively, 2 and 3 days younger than the average age for Stage X, and the estimated fertilization age of both is comparable to Stage VIII. The variability in age in Stage XI is particularly narrow, which is due, at least in part, to the fact that all the embryos are the result of single matings. Age and size are also relatively well correlated. There are two exceptions: specimen A66-88 (II) is small and A65-139 is large. The variation in greatest length in Stages IX, X, and XI is 1.3, 1.3, and 2.0 mm, respectively, which is similar to that found in the succeeding stages; however, since the total length is greater for the later stages, the degree of variation is greater than in the succeeding stages. The variation in dorsal flexure and the use of different fixatives may, in part, account for the variability. Streeter (1942) has discussed the various factors which contribute to the degree of the dorsal flexure. One reason given for this flexure occurring is the extensive handling that is required for processing. In one case (A66-14), an embryo with an extreme dorsal flexure was observed immediately after dissection of the unfixed endometrial sac. At least in this instance the dorsal flexure was present before processing was started. Embryo A66-88 represents an unusual case in that it is one of a set of twins which may also account for its smaller size. The other embryo is 7 days younger developmentally (Hendrickx, Houston, and Kraemer 1968).

Of the 10 single matings, 6 of them occurred on or within 1 day of the optimal mating time (the 3d day preceding deturgescence). Four of them occurred earlier, one (A68-122, X) on day 5, two (A65-174, IX, and A64-119, XI) on day 6, and one (A66-30, IX) on day 7 preceding deturgescence. The estimated fertilization age of the embryos from the early matings corresponds more closely with the developmental characteristics for the respective stages than does the insemination age.

TABLE 5.1

EMBRYOS OF STAGES IX–XI

	INSEM. AGE (Days)	EFA (Days)	MIN. AGE (Days)	GREATEST LENGTH (mm)	SOMITES	MATING	
						Single	Cont.
Stage IX							
A65-149	24	22	20	0.77	...		×
A65-150	...	25	24	0.80	...		×
A66-30	26	24	24	0.71	...	×	
A65-174	25	23	23	1.3	3	×	
A65-172	...	28	27	1.9	3		×
Stage X							
A68-128	24	24	24	2.7	4	×	
A65-142	...	24	23	2.3	5		×
A65-176	...	22	21	2.5	5		×
A65-139	...	24	23	3.3	6		×
A66-88 (II)	27	26	25	2.0	6	×	
A65-179	...	23	23	2.5	8		×
A68-122	27	25	25	3.0	9	×	
A65-157	...	25	24	2.9	11		×
Stage XI							
A65-219	28	27	26	3.3	15	×	
A66-14	27	27	26	3.4	15	×	
A67-133	27	27	27	3.1	15	×	
A65-201	26	25	24	4.4	16	×	
A64-119	29	27	27	4.3	19	×	

The diameter of the chorion no longer provides a good staging characteristic because of the small increments in size, consequently it will not be considered (see Appendix, fig. A. 17 for chorionic diameter of Stages IX–XXIII).

II. External Characteristics

The presomite embryos of Stage IX are similar in shape to those of Stage VIII, all having curvatures in the transverse and longitudinal planes (fig. 5.1*A–C*). In dorsal view the embryonic shield appears as a pear-shaped disc. The body stalk varies in size between the presomite and early somite embryos, becoming more condensed in the more advanced specimens (fig. 5.1).

The early somite embryo is elongated and is expanding into the amniotic cavity (fig. 5.1*D*). The yolk sac is beginning to constrict at the ventral surface of the embryo. The head fold has formed and the tail fold is in the early stages of formation (fig. 5.6*A–C*). Two prominent elevations, the neural folds, form over the cranial two-thirds of the embryo (fig. 5.1*D*). Caudally, the neural folds fade out in the region of the primitive pit. The cranial flexure appears in the presumptive midbrain region. Somites appear as condensations of the paraxial mesoderm lateral to the neural folds near the cephalic end of the notochord. Each somite is set off by an intersomitic groove, a slightly depressed area between each somite (fig. 5.1*D*). A slight dorsal flexure, or kink, develops in the somite region.

Stage X embryos are characterized by the development of 4 to 12 somites and the closure of the neural folds (figs. 5.2, 5.3). The neural folds fuse first between the 4th and 8th somites to form the neural tube. The optic primordia appear as bulges in the cranialmost region of the neural folds and the otic placodes are recognized as slight platelike depressions of the surface ectoderm on each side of the hindbrain (fig. 5.2*C*). The formation of the neural tube initiates a change in the external form from the relatively flat embryonic plate to a somewhat cylindrical, elongated structure.

The mandibular and hyoid arches appear as the 7th and 8th somites develop, respectively. The heart forms a bulge which is distinguishable on the external surface in all but the youngest embryos. The cranial flexure is prominent, and the dorsal flexure is variable in its degree but occurs in most of the embryos.

The tail fold is formed and the embryo is now elevated above the yolk sac. The dorsal flexure is present to some degree in all embryos of Stage XI and there is still a broad communication between the vitelline sac, foregut, and hindgut. The mandibular and hyoid arches are well defined. The invaginating otic placode is becoming more distinct (fig. 5.4*D*).

The neural tube now extends the full length of the embryo, resulting in an elongated cylindrical body. The cranial (anterior) neuropore is a distinct landmark when observed in dorsal or frontal view. In the more transparent specimens, the

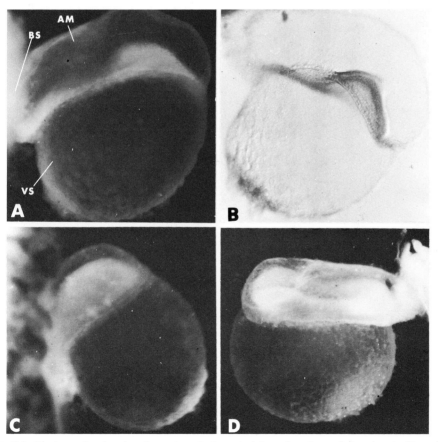

Fig. 5.1. Photographs showing the external characteristics of embryos belonging to Stage IX. *A*, right lateral view of a 22-day presomite embryo. (×65) *B*, dorsal view of a 22-day presomite embryo. (×65) *C*, right lateral view of a 24-day presomite embryo. (×60) *D*, left dorsolateral view of a 3-somite embryo. (×20) *A–C*, the side-to-side and craniocaudal curvature of the embryonic disc appears in all three presomite specimens. The embryo is enveloped in the amnion (*AM*) dorsally and the vitelline sac (*VS*) ventrally. The heavy line at the juncture of these two membranes is the edge of the embryo. The body stalk (*BS*) in *A* is more condensed than that shown in *C* where it extends over the embryo dorsally.

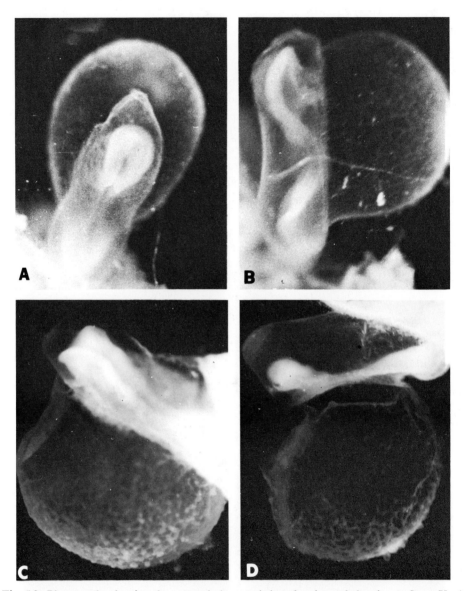

Fig. 5.2. Photographs showing the external characteristics of embryos belonging to Stage X. *A*, dorsal view of a 4-somite embryo. (×17) *B*, right view of the same embryo. (×20) *C*, left dorso-lateral view of a 5-somite embryo. (×20) *D*, left view of a 7-somite embryo. (×17)

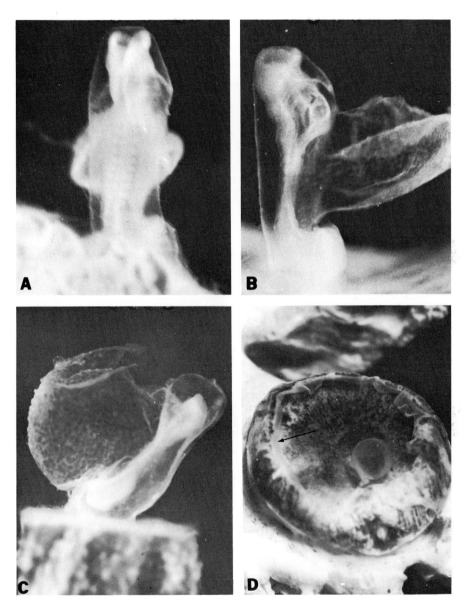

Fig. 5.3. Photographs showing the external characteristics of embryos belonging to Stage X. *A, B*, dorsal and right views of a 9-somite embryo. (×20) *C*, left view of a 6-somite embryo. (×13) *D*, 6-somite embryo and placenta. (×3) The membranous chorion has been cut away *(arrow)*.

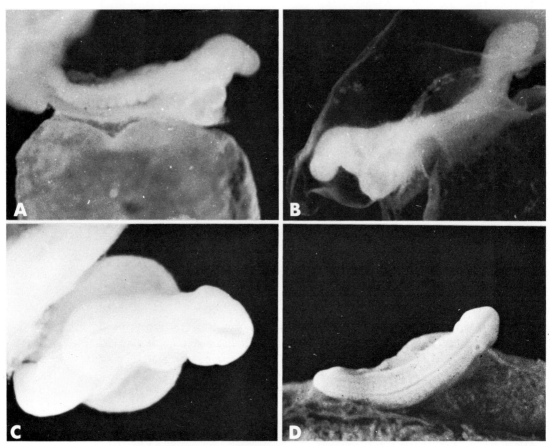

Fig. 5.4. Photographs showing the external characteristics of embryos belonging to Stage XI. A, B, right and left views of a 15-somite embryo. (×23) C, dorsal view of a 15-somite embryo. (×20) D, dorsal view of a 19-somite embryo with the caudal end lying parallel to the placenta. (×17)

S-shaped loop of the developing heart is visible. The cervical flexure is also detectable.

III. Internal Characteristics

Stained, serial sections of the embryos were examined microscopically. The structures discussed below were examined for their level of development. Each specimen was placed subsequently in a particular stage based on the combined level of development of these structures.

A. *Nervous System* (Figs. 5.5 to 5.10)

The neural folds appear early in Stage IX (fig. 5.5*C*). In the older embryo there are two elevated ridges which flank the longitudinal axis of the embryo. Cranial to the cranial flexure, the neural groove is somewhat shallow but becomes deeper and V-shaped in slightly more caudal regions (figs. 5.6; 5.7*C–I*). Caudal to the somite region the V-shaped groove becomes a shallow depression (fig. 5.7–*L*). The neural plate disappears at the primitive node.

The neural folds approach each other early in Stage X and begin to close over the neural groove in embryos with 8 somites. Closure first takes place between somites 4 and 8, and progresses cranially and caudally at nearly the same rate. In embryos with 11 somites the neural tube is formed cranial to the first somite and just caudal to the last somite. The primary brain regions—the prosencephalon, mesencephalon, and rhombencephalon—also appear early in Stage X. Sulci in the lateral walls of the neural folds separate the respective brain vesicles in the 8-somite embryo, and by 11 somites the prosencephalon has expanded. Neural crest cells, representing the trigeminal and facioacoustic primordia, become apparent early in this stage (figs. 5.8; 5.9*E–G*). They consist of a small collection of cells extending from the juncture of the neural fold with the surface ectoderm. The neural crest cells contain some cytoplasm and have large, darkly stained, spherical- to ellipsoidal-shaped nuclei. Although they are more closely packed than the surrounding mesenchymal cells, they are not separated by a distinct boundary (Gasser and Hendrickx 1967; 1969).

A craniocaudal gradient of development is reflected in the closure of the cranial neuropore in Stage XI (the caudal neuropore does not close until Stage XII). The neural tube is closed throughout the primary brain vesicle region. By 16 somites the lips of the neural tube are closed midway through the optic vesicle region. By the end of this stage, fusion of the neural folds is complete and fusion of the overlying skin ectoderm is almost complete. Midway through this stage, the 3 primary brain vesicles are distinct and the primary rhombomeres are formed (fig. 5.10*A–F*). By 19 somites, 5 brain vesicles are formed and the primordium of cranial nerve IX is present.

B. *Eye (Optic Vesicle)* (Figs. 5.9*A;* 5.10*C–F*)

The optic primordium is first evident in Stage X when the neural tube is just beginning to close. It first appears as a slight ventrolateral bulging of the prosencephalon (fig. 5.9*A*).

Evagination continues in Stage XI in the lateral wall of the forebrain forming the optic sulcus. The optic vesicle soon develops as a large out-pocketing directed laterally (fig. 5.10*C–F*). Midway through the stage the optic cavity appears, first in the form of a slit, which subsequently becomes a cavity that communicates with the prosocele. The vesicle is separated from the surface ectoderm by a thin layer

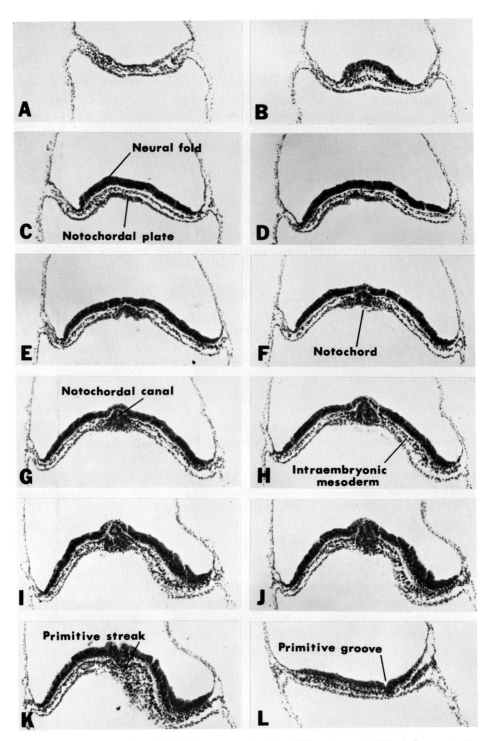

Fig. 5.5 Photomicrographs of a 22-day, presomite, Stage IX embryo. (×50) *A–L,* representative, transverse sections in serial order through the neural fold, notochordal plate, notochord, notochordal canal, intraembryonic membrane, primitive streak, and primitive groove.

of mesoderm. It is set off from the lateral surface of the brain by cranial- and caudal-limiting sulci. The cranial-limiting sulcus is only faintly visible since it is located in a region where the vesicle merges with the brain wall.

C. *Ear* (*Otic Vesicle*) (Figs. 5.9H; 5.10B–C)

The otic placode is a readily detectable structure at Stage X and is present in all of the 5- and 6-somite embryos of this stage. It provides an excellent landmark for early brain, cranial nerve, and pharyngeal arch development. The placode first appears as a plate of slightly thickened epithelium which begins to invaginate in the older specimens.

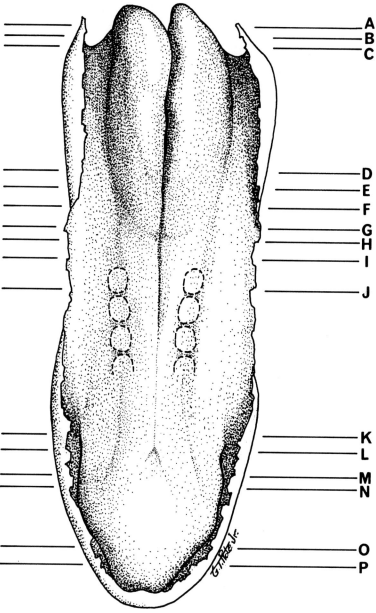

Fig. 5.6. Drawing taken from a reconstruction of a 23-day, 3-somite, Stage IX embryo, dorsal view. (×85) Lines *A–P* indicate the level and plane of each section shown in fig. 5.7.

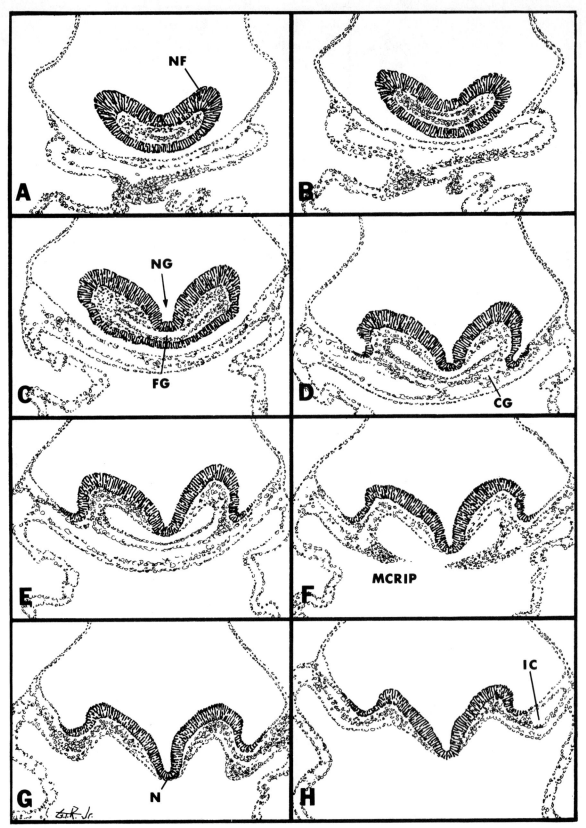

Fig. 5.7. Drawings of representative, transverse sections in serial order of the Stage IX embryo shown in fig. 5.6. (×120) Allantois, *AL;* cardiogenic mesoderm, *CG;* cloacal membrane, *CM;* foregut, *FG;* hindgut, *HG;* intraembryonic coelom, *IC;* margin of the cranial intestinal portal, *MCRIP;* margin of the caudal intestinal portal, *MCAIP;* notochord, *N;* neural fold, *NF;* neural groove, *NG;* primitive streak, *PS;* 1st somite, *S1.*

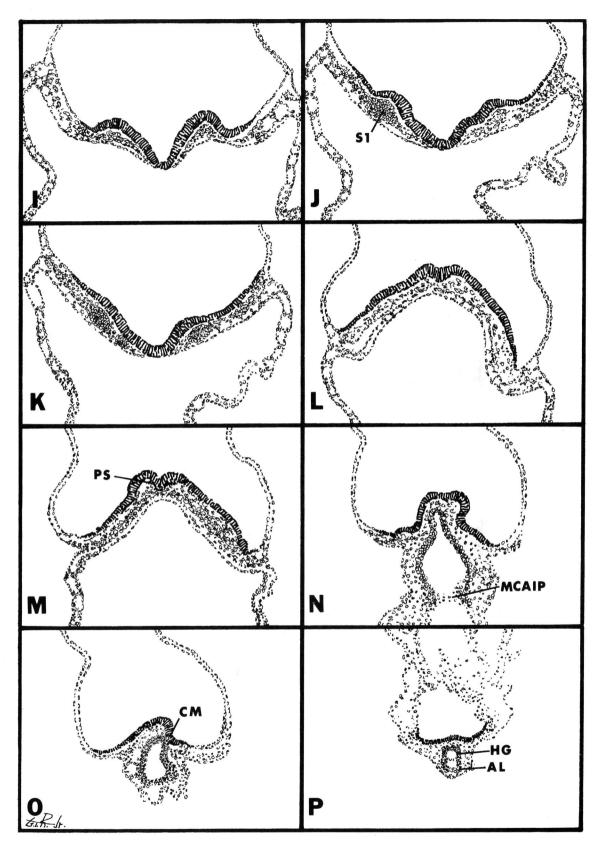

Fig. 5.7—*Continued*

In the 13-somite Stage XI embryos the placode is slightly invaginated and consists of an epithelial plate, 2 or 3 cells thick. The nuclei are eccentrically placed near mesenchymal tissue. By 16 somites it has invaginated further and becomes slightly thicker. The placode is over half closed at 20 somites, but the opening to the outside, the otic pit, is still apparent. The junction of the placode with the skin ectoderm is distinct (fig. 5.10*C*).

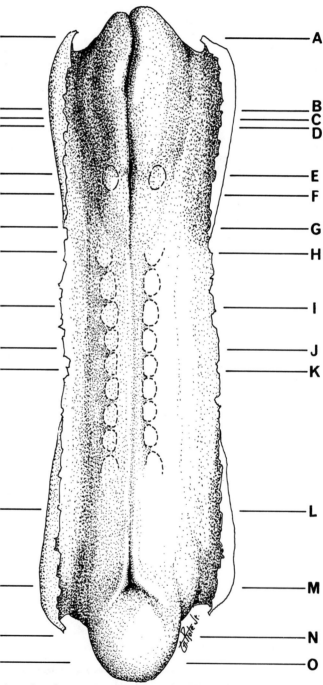

Fig. 5.8. Drawing taken from a reconstruction of a 23-day, 8-somite, Stage X embryo. (×70) Lines *A–O* indicate the level and plane of the sections shown in fig. 5.9.

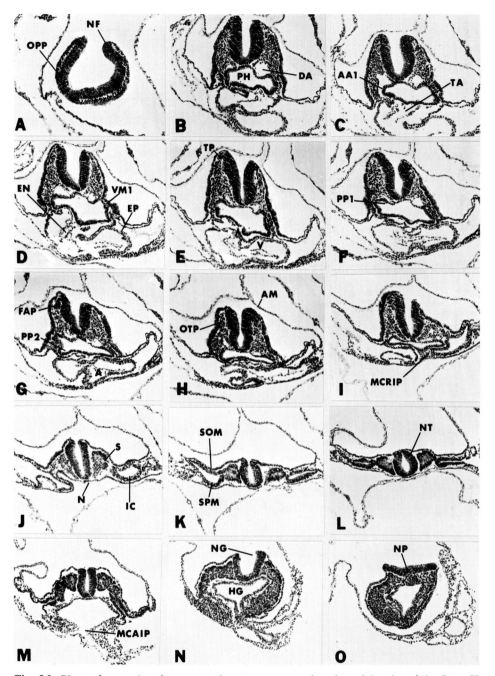

Fig. 5.9. Photomicrographs of representative, transverse sections in serial order of the Stage X embryo shown in fig. 5.8. (×65) Atrium, *A;* 1st aortic arch, *AA1;* amnion, *AM;* dorsal aorta, *DA;* endocardium, *EN;* epicardium, *EP;* facioacoustic primordium, *FAP;* hindgut, *HG;* intraembryonic coelom, *IC;* margin of the cranial intestinal portal, *MCAIP;* margin of the caudal intestinal portal, *MCAIP;* notochord, *N;* neural fold, *NF;* neural groove, *NG;* neural plate, *NP;* neural tube, *NT;* optic primordium, *OPP;* otic placode, *OTP;* pharynx, *PH;* 1st pharyngeal pouch, *PP1;* 2d pharyngeal pouch, *PP2;* somite, *S;* somatic mesoderm, *SOM;* splanchnic mesoderm *SPM;* truncus arteriosus, *TA;* trigeminal primoridum, *TP;* ventricle, *V;* 1st visceral membrane, *VM1.*

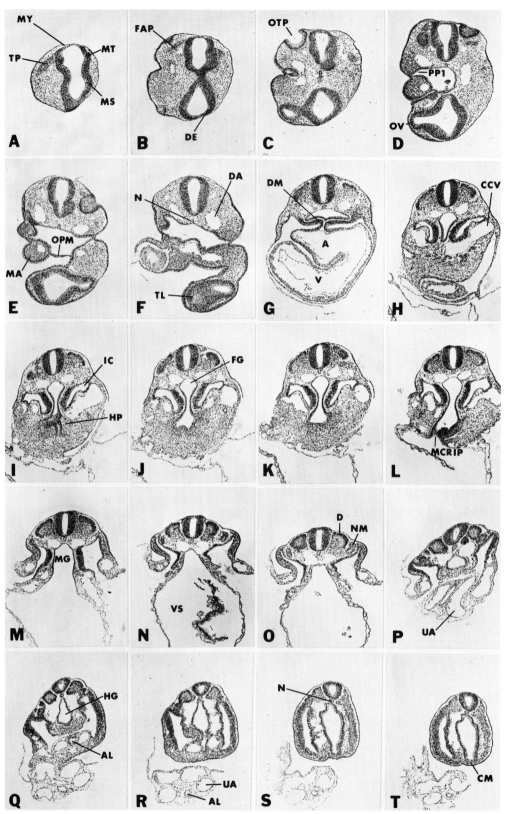

Fig. 5.10. Photomicrographs of representative, transverse sections in serial order of the 27-day, 19-somite, Stage XI embryo shown in fig. 5.4D. (×35) Atrium, *A;* allantois, *AL;* common cardinal vein, *CCV;* cloacal membrane, *CM;* dermo-myotome, *D;* dorsal aorta, *DA;* diencephalon, *DE;* dorsal mesocardium, *DM;* facioacoustic primordium, *FAP;* foregut, *FG;* hindgut, *HG;* hepatic primordium, *HP;* intraembryonic coelom, *IC;* mandibular arch, *MA;* margin of the cranial intestinal portal, *MCRIP;* midgut, *MG;* mesencephalon, *MS;* metencephalon, *MT;* myelencephalon, *MY;* notochord, *N;* nephrogenic mesoderm, *NM;* otic pit, *OTP;* oro-pharyngeal membrane, *OPM;* optic vesicle, *OV;* 1st pharyngeal pouch, *PPI;* telencephalon, *TL;* trigeminal primordium, *TP;* umbilical artery, *UA;* ventricle, *V;* vitelline sac, *VS.*

D. *Heart* (Figs. 5.7D; 5.9B–I; 5.10F–H)

The cardiogenic area appears as a plexiform mass of mesenchyme ventral to the foregut and cranial to the cranial intestinal portal at Stage IX.

By Stage X, in embryos with 5 to 6 pairs of somites the two endocardial tubes are fused at the aortic bulb, bulbus cordis, and ventricular region. At 8 somites, fusion through the ventricular area is complete but the atria are still paired. The ventricular tube gradually changes from a relatively straight structure into an S-shaped loop. The enveloping myocardial wall develops concurrently with the formation of the S-shaped loop in the older embryos. The 1st aortic arch is well formed and the 2d aortic arch is appearing by the end of Stage X (fig. 5.9C).

The increasing complexity of the developing heart makes it a less valuable staging characteristic and will not be used beyond Stage XI. The S-shaped loop which appeared in Stage X is very prominent at Stage XI. The aortic bulb, bulbus cordis, ventricle, atrium, and sinus venosus are readily recognizable. The endocardium and myocardium are separate layers. The dorsal mesocardium persists (fig. 5.10G). The dorsal aortae are still paired, and 3 aortic arches are present, the 3d being incompletely developed.

E. *Endodermal Derivatives* (Figs. 5.7, 5.9, 5.10)

With the formation of the headfold in Stage IX, the dorsal part of the vitelline sac is incorporated into the embryo, forming the foregut cranially. Likewise, a short hindgut forms caudally with the development of the tail fold. The midgut is in broad communication with the vitelline sac through the vitello-intestinal isthmus. The cranial and caudal intestinal portals are wide, allowing for free communication between the fore, mid, and hindgut (figs. 5.7F, N; 5.9I, M; 5.10L). A slight lateral expansion of the foregut marks the early primordia of the 1st pharyngeal pouch bordered by the primordia of the 1st and 2d visceral arches. The oro-pharyngeal membrane is forming but a stomodeum is not apparent. The allantois first appears as a short cul-de-sac which projects into the body stalk just ventral to the hindgut. The allantois is longer than the hindgut.

The oro-pharyngeal membrane and stomodeum are well defined at Stage X. The 1st visceral arch (mandibular) is evident and the 2d visceral arch (hyoid) is indicated by the formation of the 2d pharyngeal pouch. The 1st visceral membrane is eminent (fig. 5.9D). The margins of the cranial and caudal intestinal portals mark the cranial and caudal borders, respectively, of the vitello-intestinal isthmus. The hindgut has elongated but otherwise remains unchanged.

By Stage XI the oro-pharyngeal membrane is reduced to a single cell layer but remains intact (fig. 5.10E). The mandibular arch is clearly outlined by the oral groove cranially and the 1st visceral groove caudally (fig. 5.10E). It has expanded from the previous stage and comprises the greater part of the pharyngeal complex. The hyoid arch has become better defined since it is now outlined by the 1st and 2d visceral grooves. A condensation of mesenchyme caudal to the 2d arch represents the 3d visceral arch. A shallow groove caudal to the 3d arch represents the 3d visceral groove. Three pharyngeal pouches and two visceral membranes are present. The thyroid primordium is represented by cellular processes from the pharyngeal floor into the mesenchyme. The primordium of the liver is a ventral outgrowth of the foregut just cranial to the cranial intestinal portal (fig. 5.10H–K). The hindgut has elongated, but otherwise shows no changes that are readily applicable to classifying embryos; so it will not be considered beyond this stage.

F. *Notochord and Primitive Streak* (Figs. 5.5, 5.7, 5.9, 5.10)

1. *Notochord.* The notochord becomes prominent in the presomite and early somite embryos of Stage IX. The notochordal (neurenteric) canal is well formed in the presomite embryos and extends the full length of the notochord (fig. 5.5*E–K*). It communicates with the vitelline cavity at its cranial end (fig. 5.5*E*) and with the amniotic cavity at its caudal end (fig. 5.5*K*). The columnar epithelial cells comprising the notochord are oriented around the central canal. Where the canal communicates with the vitelline cavity, the notochord is a semilunar structure (fig. 5.5*E*). At points along the notochord the cells lose their orderly arrangement and obliterate the notochordal canal. The opening of the notochordal canal into the primitive streak is a small slit which is not found consistently. In the Stage IX embryos possessing somites, the notochord is embedded in the endoderm and is in close contact to the ventral surface of the neural plate (fig. 5.7*G*). It is distinguished from the endoderm by the columnar appearance of its cells. The notochord extends cranially slightly beyond the head fold and caudally to the primitive streak.

By Stage X the notochord consists of a bar of cells that extends from the cranial end of the foregut to the primitive streak with which it merges (fig. 5.9*J*). In all the specimens of this stage the notochord is either embedded or closely related to the dorsal endoderm over most of its length. It forms a separate rodlike structure for several sections in the older specimens. By Stage XI the notochord is cordlike and is separated from the endoderm except in the pharyngeal region (fig. 5.10*F*).

2. *Primitive streak.* The primitive streak remains conspicuous in the presomite embryos of Stage IX (fig. 5.5*K*). As the somites differentiate it is less evident but retains its characteristic shape in the tail region. By Stage X it is rapidly disappearing. In the youngest embryos it comprises approximately 25% of the total length but decreases to about 12% in the older specimens. The primitive streak is not apparent at Stage XI.

G. *Intraembryonic Coelom* (Figs. 5.5*I;* 5.7*H;* 5.9*D–M;* 5.10*G–T*)

The small, isolated clefts in the intraembryonic (lateral plate) mesoderm begin to coalesce in the presomite embryos of Stage IX. Such coalescence is widespread in the embryos possessing somites. The cavitation continues in Stage X where the intraembryonic coelom divides the lateral-plate mesoderm into a dorsal somatic layer and a ventral splanchnic layer (fig. 5.9*K*). By Stage XI the intraembryonic coelom is well formed.

References

Gasser, R. F., and A. G. Hendrickx 1967. The development of the facial nerve in baboon embryos (*Papio* sp.). *J. Comp. Neurol.* 129:203–18.

———— 1969. The development of the trigeminal nerve in baboon embryos (*Papio* sp.). *J. Comp. Neurol.* 136:159–82.

Hendrickx, A. G.; M. L. Houston; and D. C. Kraemer 1968. Observations on twin baboon embryos (*Papio* sp.). *Anat. Rec.* 160:181–86.

Heuser, C. H., and G. W. Corner 1957. Developmental horizons in human embryos. Description of age group X, 4 to 12 somites. *Contrib. Embryol., Carneg. Inst.* 36:29–39.

O'Rahilly, R. 1963. The early development of the otic vesicle in staged human embryos. *J. Embryol. Exp. Morph.* 11:741–55.

——— 1966. The early development of the eye in staged human embryos. *Contrib. Embryol., Carneg. Inst.* 38:1–42.

Schuster, G. 1965. Untersuchungen an einem Embryo von *Papio doguera* Pucheran 7-Somiten-Stadium). *Anat. Anz.* 117:447–75.

Streeter, G. L. 1942. Developmental horizons in human embryos: Description of age group XI, 13 to 20 somites and age group XII, 21 to 29 somites. *Contrib. Embryol., Carneg. Inst.* 30:211–45.

6

Description of Stages, XII, XIII, and XIV

Andrew G. Hendrickx

I. Age and Size

STAGE XII—21–29 paired somites; EFA, 28 ± 1 days; CR length, 3–4.5 mm.
STAGE XIII—EFA, 29 ± 1 days; CR length, 4.5–6 mm.
STAGE XIV—EFA, 30 ± 1 days; CR length, 6–7 mm.

Data on the age and size of the embryos of Stages XII, XIII, and XIV are given in table 6.1. There is a close correlation between the estimated fertilization age and the stage of morphological development in all but two Stage XIV specimens (A68-98 and A65-123). The variability in age is quite narrow (1 day in Stage XII, 4 days in Stage XIV). Specimen A68-98 is from a continuous mating and has an estimated fertilization age which is equal to that of Stage XII embryos. The other nonconforming specimens, A65-123, is from a single mating and is one of the most advanced embryos of Stage XIV with an age that is similar to the embryos of Stage XV.

The age and size of one of the embryos (C 688, 30 days, CR 4.8 mm, 34 somites) studied by Gilbert and Heuser (1954) are similar to our specimens belonging to Stage XII. Another embryo which they described (C 696, 34 days, CR 6.6 mm, 40 somites) was estimated to be 2 days older than our specimens belonging to Stage XIV but is similar in size. Their method of estimating age was based on the onset of deturgescence and is less specific than our method since they did not give the estimated day of fertilization.

Age and size are well correlated with the exceptions of embryos A65-102 (XII) and A64-102 (XIV), both of which are curved extensively. The variation in crown-rump length in Stages XII and XIV is 1.9 and 2.4 mm, respectively. This is consistent with succeeding stages and is within the range reported for human embryos (Streeter 1942, 1945). All 6 of the single matings occurred on or within 1 day of the optimal mating time (the 3d day preceding deturgescence).

II. External Characteristics

In addition to the 21 to 29 pairs of somites, the size and C-shaped curvature of Stage XII embryos makes them distinguishable from embryos of the previous stage (fig. 6.1A, B, F). The dorsal flexure common to the preceding stage is no longer observed. There are 3 well-defined visceral arches separated by 3 visceral grooves.

87

A slight depression, the cervical sinus, appears in the caudal portion of the arch region. The location of the 4th visceral arch and groove are barely visible. The oral fossa is large and opened wide (fig. 6.1C). The 3 primary brain vesicles are visible from the exterior (fig. 6.1F). The thin roof of the rhombencephalon is faintly visible. The optic vesicles extend laterally and, in profile, appear as opaque circles with transparent centers. The caudal neuropore is either in the process of closing or is closed. The otocyst still opens on the surface, but in the older specimens the opening is seen only in sections. The appendicular ridge is distinct, with the primordium of the forelimb bud becoming apparent at the level of somites 8 to 13 (fig. 6.1D). The heart bulge is very prominent. The head, lateral body, and tail folds have developed rapidly over the previous stages. The communication between

TABLE 6.1

EMBRYOS OF STAGES XII–XIV

	INSEM. AGE (Days)	EFA (Days)	MIN. AGE (Days)	CROWN-RUMP LENGTH (mm)	SOMITES	MATING	
						Single	Cont.
Stage XII							
A65-133	. . .	28	27	3.4	21		×
A64-103	35	28	27	4.5	22		×
A65-102	30	29	28	2.6	26	×	
A65-132	. . .	28	27	4.3	26		×
A68-274	30	29	28	3.0	26	×	
Stage XIII							
A66-41	29	29	28	5.6	31	×	
Stage XIV							
A68-102	. . .	29	28	6.7	32		×
A65-196	38	30	29	5.8	33		×
A64-83	31	30	29	6.0	34	×	
A65-160	30	30	29	6.0	35	×	
A64-102	40	31	30	4.3	36		×
A68-98	. . .	28	27	6.6	36		×
A65-123	32	32	31	6.3	. . .	×	
A66-35	30	29	28	5.8	. . .	×	

the midgut and the vitelline sac, the vitelline duct, is broad in the youngest embryos but is reduced in size in the older embryos. The elongating tail is blunt, rounded, untapered, and directed cranially (fig. 6.1A, B).

The only embryo representing Stage XIII has 31 pairs of somites (fig. 6.2A). The curvature of the trunk is more extensive and there is a flattening in the upper cervical region as the cervical flexure develops. The *Nackengrube* of His, a depression in the lower cervical region, is apparent for the first time. The sacral flexure is further developed. The cervical sinus has receded, but the 4 visceral arches and grooves remain visible (fig. 6.3, *XIII*). The 1st visceral arch is subdividing into maxillary and mandibular processes. The telencephalon expands slightly and the dorsal wall of the rhombencephalon becomes thinner. The otocyst is completely closed and is separating from the overlying ectoderm. The tip of the forelimb bud is beginning to curve ventrally and the primordium of the hindlimb bud appears. The amnion ensheaths the body stalk, which is reduced in diameter.

The cervical flexure is more apparent in Stage XIV since the cervical portion of the trunk is flattened. The *Nackengrube* is also better defined (fig. 6.2C, D). The rostral portion of the embryo lies near or against the heart (fig. 6.2B). The cerebral hemispheres and mesenchephalon can be identified and the roof of the myelencephalon is transparent (fig. 6.2C, D). Changes in the other brain vesicles are not apparent externally. The primary head vein is easily recognized in cleared

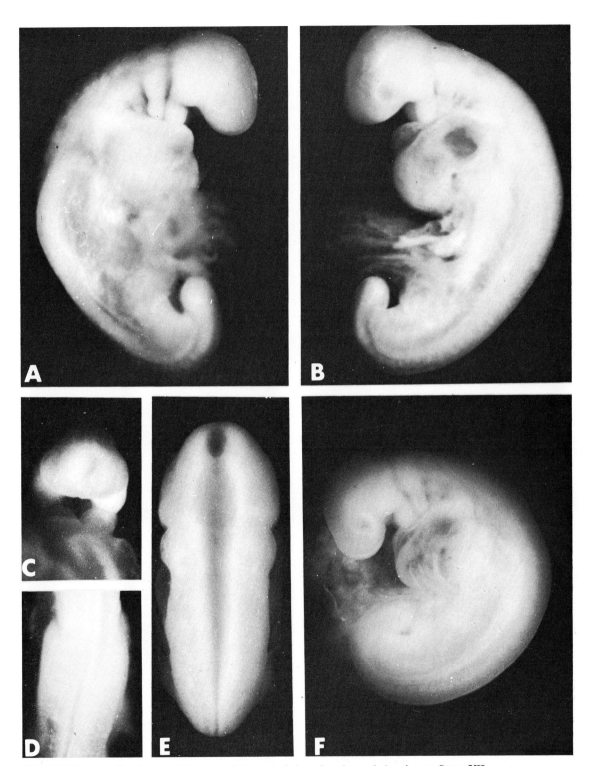

Fig. 6.1. Photographs showing the external characteristics of embryos belonging to Stage XII. *A*, right view. *B*, left view. *C*, frontal view showing the oral cavity. *D*, dorsal view showing the forelimb bud primordia. *E*, dorsal view showing contours and rhombic grooves of the hindbrain. *F*, left view of a younger embryo belonging to Stage XII. (*A–D, F,* ×20; *E,* ×35)

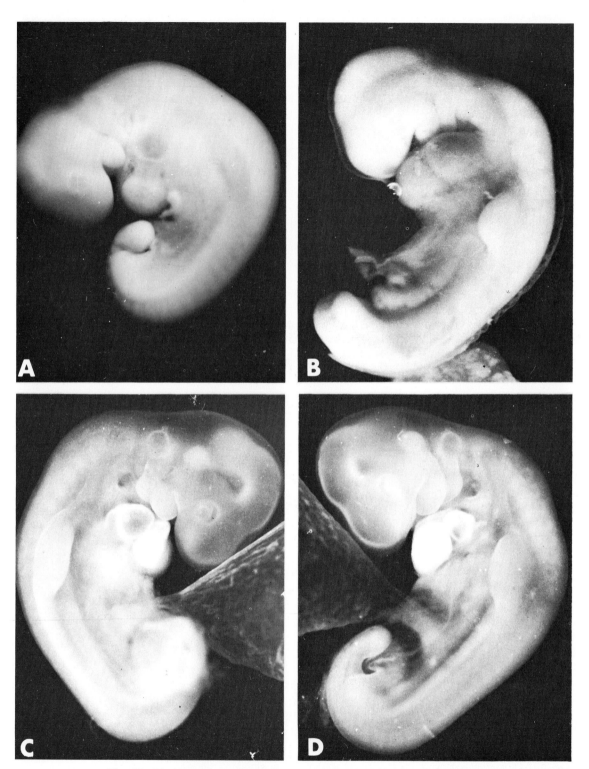

Fig. 6.2. Photographs showing the external characteristics of embryos belonging to Stage XIII and XIV. *A*, left view of a Stage XIII embryo. (×13) The forelimb buds are well defined and the hindlimb buds are becoming visible. *B*, left view of a Stage XIV embryo. (×14) *C, D*, right and left views of a Stage XIV embryo. (×13)

specimens. The eye is becoming rounded and the lens vesicle is indented but still opens to the surface. The otocyst is completely separated from the outside ectoderm and appears as a clear vesicle dorsal to the 2d visceral arch. The endolymphatic duct can be seen as a conical elevation of the dorsalmost portion of the otocyst. In the oldest embryo a shallow olfactory pit is visible adjacent to the expanding maxillary process. The maxillary and mandibular processes of the 1st visceral arch and the 2d visceral arch are prominent (fig. 6.2*C, D* and fig. 6.3, *XIV*). The deep cervical sinus conceals the 3d visceral arch. The trigeminal ganglion is a large dense area dorsal to the 1st arch. The facioacoustic primordium is smaller and less defined at the cranial aspect of the otocyst. The forelimb buds, located opposite somites 7 to 13, are slightly elongated and curve ventromedially. In the older specimens they are more tapered peripherally. The hindlimb buds are small swellings at the level of somites 27 to 30 and become finlike in the older specimens. The tail is longer with an expanded and knoblike tip.

III. Internal Characteristics

Stained, serial sections of the embryos were examined microscopically, and the level of development of the structures discussed below was determined. Each specimen was placed subsequently in a particular stage based on the combined level of development of these structures. The structures examined were the developing eye, ear, nose, hypophysis, and certain endodermal derivatives including the thyroid, trachea, esophagus, lungs, liver, and pancreas.

A. *Eye (Optic Cup and Lens Vesicle)* (Fig. 6.4)

The optic vesicle first appeared in Stage XI as an evagination of the forebrain. In Stages XII, XIII, and XIV, it differentiates into a cup and the lens vesicle develops from the surface ectoderm. By Stage XII the optic vesicle and cavity are enlarged. The optic vesicle is separated from the surface ectoderm by only a loose mesenchymal sheath (fig. 6.4, *XII-B*).

The primordium of the lens first appears and the optic cup begins to form in Stage XIII (fig. 6.4, *XIII*). The optic vesicle is enlarged and makes contact with the surface ectoderm (fig. 6.4, *XIII-B*). The ventrolateral wall of the vesicle is thickened but not indented and contains 5 to 7 rows of nuclei. The other walls of the vesicle are unchanged. The lens begins as a placode or thickening of the surface ectoderm at the point where the optic vesicle makes contact with the overlying ectoderm. The placode area is disc shaped and contains 2 to 3 rows of oval nuclei in contrast to its periphery that is a single layer of epithelium.

In Stage XIV both the optic cup and the lens placode invaginated (fig. 6.4, *XIV*). The double-walled optic cup is composed of a thick inner layer and a thin outer layer. The inner layer is about three times thicker than the outer layer and will become the neural layer of the retina. It contains 5 or 6 rows of nuclei. The outer layer is several layers thick and contains no pigment, but it will become the pigmented layer of the retina. The rim of the optic cup is incomplete ventrally, and the resulting gap constitutes the optic fissure. This fissure extends to the base of the optic cup but not into the optic stalk. The cavity of the optic cup communicates with that of the forebrain through a short optic stalk. The lens placode is indented and a distinct lens pit is formed in the older specimens (fig. 6.4, *XIV-B*). The lens pit often contains clumps of cell remnants and communicates freely with the exterior by way of a wide pore.

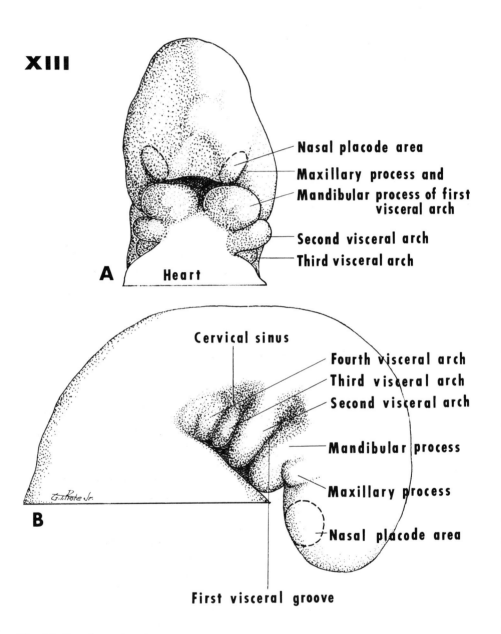

XIII

Nasal placode area

Maxillary process and

Mandibular process of first
visceral arch

Second visceral arch

Third visceral arch

A Heart

Cervical sinus

Fourth visceral arch

Third visceral arch

Second visceral arch

Mandibular process

Maxillary process

Nasal placode area

B

First visceral groove

Fig. 6.3. Drawings taken from reconstructions of the face in embryos of Stages XIII and XIV. (×35) *A*, frontal views. *B*, right views.

XIV

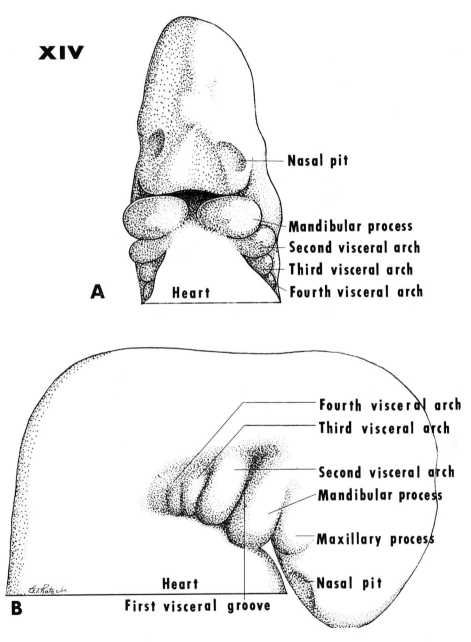

A — Nasal pit, Mandibular process, Second visceral arch, Third visceral arch, Heart, Fourth visceral arch

B — Fourth visceral arch, Third visceral arch, Second visceral arch, Mandibular process, Maxillary process, Nasal pit, Heart, First visceral groove

Fig. 6.3—*Continued*

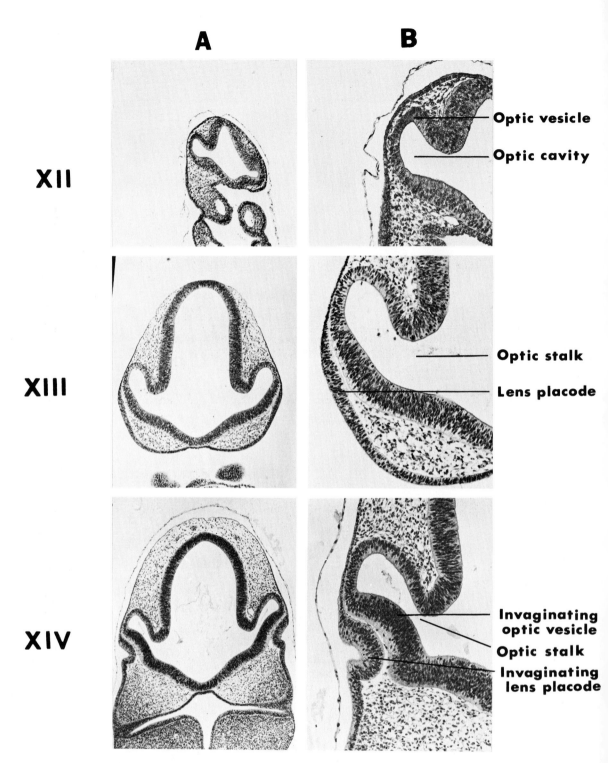

Fig. 6.4. Photomicrographs of sections of the eye near the midsection in embryos of Stages XII, XIII, and XIV. The general features of the eye and related structures are shown. (*A*, ×40; *B*, ×100)

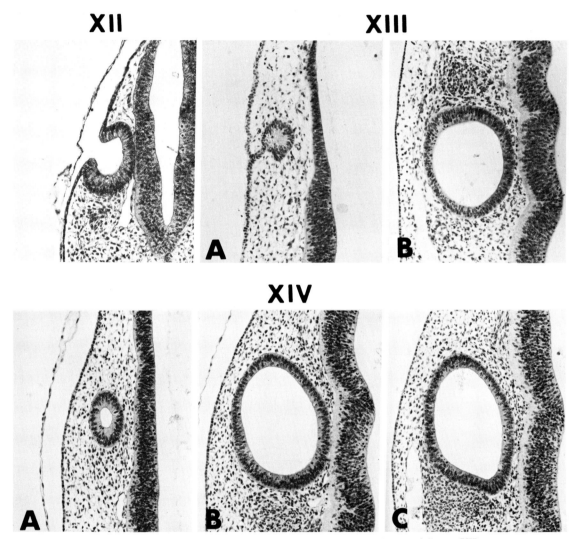

Fig. 6.5. Photomicrographs of transverse sections of the left otocyst in embryos of Stages XII, XIII, and XIV. (×100)

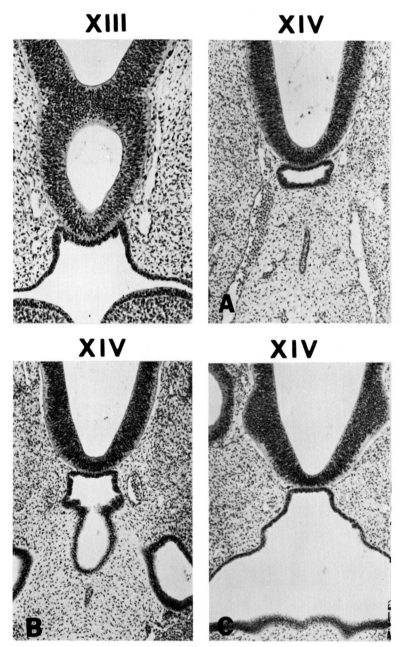

Fig. 6.6. Photomicrographs of sections of the hypophysis in the embryos of Stages XIII and XIV. ($\times$70)

B. *Inner Ear (Otocyst)* (Fig. 6.5)

The otocyst is another reliable index for determining the degree of development in baboon embryos. At Stage XII the otic pit is deeper than the previous stage and is almost closed (fig. 6.5, XII). By Stage XIII the closure of the pit is complete and the connection with the ectoderm is reduced to a short, solid cord of cells (fig. 6.5, *XIII-A*) or has disappeared. A small diverticulum, the endolymphatic duct, forms as a conical extension from the dorsomedial surface of the otocyst while the otocyst is attached to the external surface. The cells in the walls of the endolymphatic duct are less compact and contain more cytoplasm than those of the remainder of the otocyst (fig. 6.5, *XIII-A*). In Stage XIV both the otocyst and the endolymphatic duct increase in size. Differential thickening in the wall of the otocyst is apparent (fig. 6.5, *XIV-B, C*).

C. *Hypophysis (Hypophyseal Pouch)* (Fig. 6.6)

The hypophyseal or Rathke's pouch appears late in Stage XII as a small outpocketing in the roof of the stomodeum immediately cranial to the oro-pharyngeal membrane. It is in close contact with the floor of the prosencephalon and begins to grow cranially in Stages XIII and XIV. Its connection with the stomodeum narrows by Stage XIV (fig. 6.6, *XIV-B*). The neurohypophyseal primordium is not yet apparent.

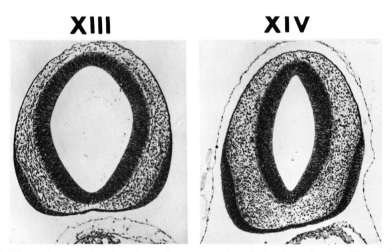

Fig. 6.7. Photomicrographs of sections of the nasal placodes in the embryos of Stages XIII and XIV. (×50)

D. *Nose (Nasal Placode and Pit)* (Figs. 6.3, 6.7)

The development of the nose is associated with the changes that produce the face and the palate. The nasal placode is evident first in Stage XII and by Stage XIII appears as an ectodermal thickening on the ventrolateral surface of the head. By Stage XIV the placode is thicker and begins to invaginate forming a nasal pit.

E. *Endodermal Derivatives* (Figs. 6.8 to 6.11)

All 4 pharyngeal pouches are present in the older embryos of Stage XII. Three pharyngeal membranes are evident in Stage XII, 4 in Stage XIII, and the ultimobranchial body appears in Stage XIV. The oro-pharyngeal membrane is ruptured by Stage XII. The thyroid diverticulum in Stage XII embryos is a slight thickening in the floor of the pharynx. By Stage XIII it consists of a bilobed primordium with a wide attachment to the floor of the pharynx. In Stage XIV embryos

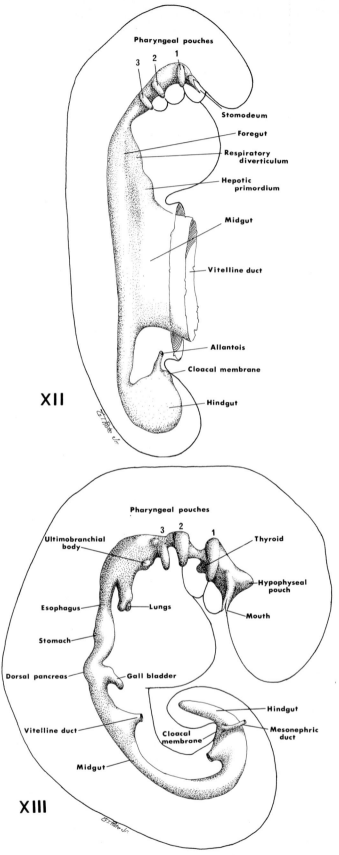

Fig. 6.8. Drawings taken from reconstructions of the alimentary and respiratory systems of embryos of Stages XII, XIII, and XIV.

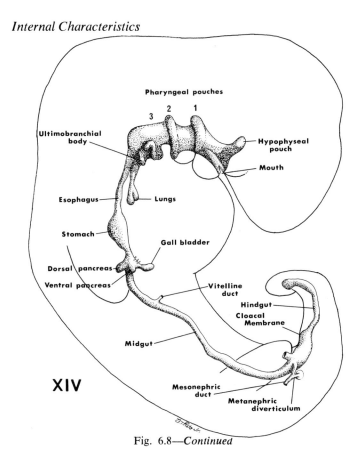

Fig. 6.8—*Continued*

the connection between the thyroid primordium and the pharynx is reduced to a stalk, the thyroglossal duct.

The respiratory groove is well formed early in Stage XII. Longitudinal grooves appear on each side of the foregut; these deepen and meet in the midline by Stage XIII, separating the primitive esophagus from the primitive trachea (fig. 6.9). Coincidental with the formation of the primitive trachea, the caudal tip of the respiratory diverticulum enlarges and then bifurcates, forming two short lung buds by Stage XIII (fig. 6.10). The lung buds elongate and expand in Stage XIV.

Both the liver and pancreas undergo notable changes in Stages XII, XIII, and XIV. By Stage XII the hepatic diverticulum contains cords of cells which extend into the septum transversum (fig. 6.11, *XII*). The number of cords increases by Stage XIII when the gall bladder is recognizable (fig. 6.11, *XIII*). The dorsal pancreas appears as a small, dorsal outgrowth of the primitive duodenum (fig. 6.11, *XIII-A*). By Stage XIV it has enlarged and is easily recognized. The ventral pancreas appears in the angle between the gut and hepatic diverticulum (figs. 6.11, *XIV-A;* 7.16).

The stomach is defined as a dilatation of the foregut in Stage XII and becomes fusiform by Stage XIII. In Stage XIV its dorsal aspect is partially rotated to the left and the midgut loop extends towards the umbilical cord. The hindgut elongates in Stages XII and XIII and extends deep into the tail by Stage XIV. The cloacal membrane remains intact.

F. *Metanephros*

The primordium of the definitive kidney first appears in Stage XIV. The metanephric diverticulum arises from the mesonephric duct near the point where it joins the cloaca (fig. 6.8, *XIII*). It is a short, hollow process that extends dorso-cranially.

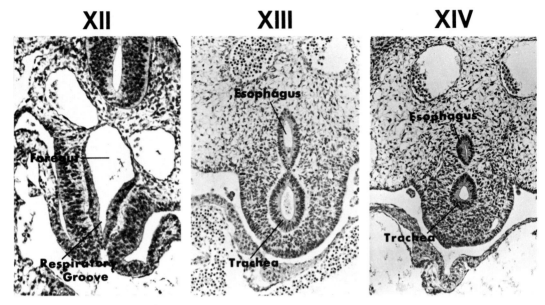

Fig. 6.9. Photomicrographs of transverse sections of the foregut and respiratory groove, and primitive esophagus and trachea at comparable levels in embryos of Stages XII, XIII, and XIV. (×85)

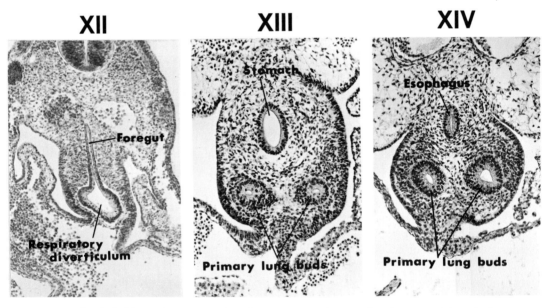

Fig. 6.10. Photomicrographs of sections of the respiratory diverticulum and lung buds at comparable levels in embryos of Stages XII, XIII, and XIV. (×85)

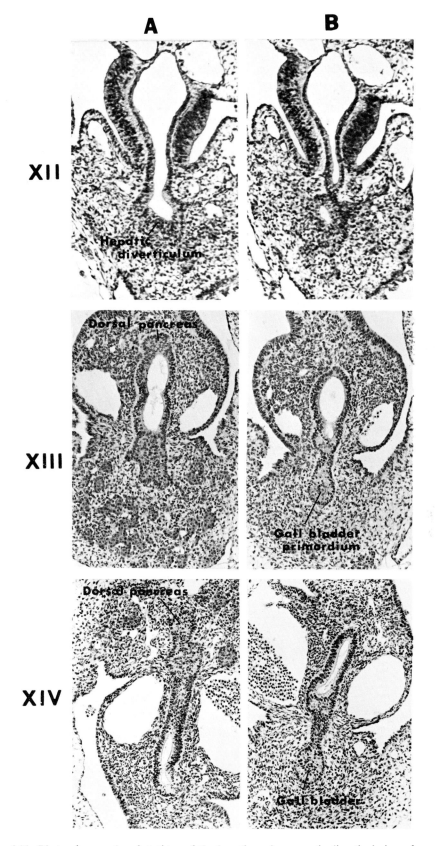

Fig. 6.11. Photomicrographs of sections of the hepatic and pancreatic diverticula in embryos of Stages XII, XIII, and XIV. (×85)

References

Gilbert, C., and C. H. Heuser 1954. Studies in the development of the baboon embryo (*Papio ursinus*). *Contrib. Embryol., Carneg. Inst.* 35:13–54.

O'Rahilly, R. 1966. The early development of the eye in staged human embryos. *Contrib. Embryol., Carneg. Inst.* 38:1–42.

Streeter, G. L. 1942. Developmental horizons in human embryos: Description of age group XI, 13 to 20 somites, and age group XII, 21 to 29 somites. *Contrib. Embryol., Carneg. Inst.* 30:211–45.

———— 1945. Developmental horizons in human embryos: Description of age group XIII, embryos about 4 or 5 millimeters long, and age group XIV, period of indentation of the lens vesicle. *Contrib. Embryol., Carneg. Inst.* 31:27–63.

7

Description of Stages XV, XVI, XVII, and XVIII

Andrew G. Hendrickx/Joe A. Bollert/Marshall L. Houston

I. Age and Size

STAGE XV—EFA, 31 ± 1 days; crown-rump length, 6–8 mm.
STAGE XVI—EFA, 33 ± 1 days; crown-rump length, 7–9 mm.
STAGE XVII—EFA, 35 ± 1 days; crown-rump length, 10–13 mm.
STAGE XVIII—EFA, 37 ± 1 days; crown-rump length, 14–17 mm.

Data on the age and size of the embryos of Stages XV through XVIII are given in table 7.1. The embryos are listed in the order of their morphological development. There is a close correlation between the estimated fertilization age and morphological development in all but 4 cases, A68-252 (XVII), A64-96 (XVIII), A65-197 (XVIII), and A65-199 (XVIII). The variability in age is quite narrow, 1 day for Stages XV and XVI, but widens to 5 and 6 days respectively for Stages XVII and XVIII. Embryo A68-252 from a continuous mating is the oldest specimen of Stage XVII, and its age is similar to embryos of Stage XIX. Embryo A64-96 (XVIII) from a continuous mating is similar in age to Stage XIX embryos; and embryos A65-199 and A65-197, both from single matings, are similar in age to embryos of Stages XVII and XXI respectively.

The variation in crown-rump length in Stages XV to XVIII is 1.2, 2.1, 1.9, and 1.9 mm respectively, which is consistent with preceding and succeeding stages and is within the range reported for human embryos (Streeter 1948). Age and size are relatively well correlated; however, there are inconsistencies. This is explained in part by the degree of curvatures as well as the use of different fixatives. Specimen A65-116 (XVI) is small and is curved more than usual. The increase in length from Stage XVI to Stage XVII is due, in part, to the straightening of the back as well as to the movement of the head up off the heart.

Of the 7 single matings, 3 of them occurred on the 3d or 4th day preceding deturgescence, 2 occurred on the 5th, 1 on the 6th, and 1 on the 9th day preceding deturgescence. Embryo A66-75 represents the only successful pregnancy in which mating occurred as early as the 9th day preceding deturgescence. The estimated fertilization age of the embryos from the early matings corresponds more closely with the developmental characteristics for the respective stages than does insemination age.

103

II. External Characteristics

The external features which characterize Stage XIV, namely the head region and the appendages, become more prominent (figs. 7.1 to 7.6). In the head region, the lens vesicles are closed and no longer communicate with the surface. The nasal placodes recede from the surface, forming large oval depressions. The visceral arches undergo gradual changes, forming the primordium of the face and auricles. The heart becomes a distinct feature and plays a major role in the shaping of the trunk. Widening of the trunk begins in the occipitocervical region and continues caudally, nearly doubling from Stage XV through Stage XVIII. The arm buds are subdivided into hand and arm-shoulder segments, and in a slightly delayed sequence, the leg buds also are subdivided into foot and leg-thigh segments.

TABLE 7.1

EMBRYOS OF STAGES XV–XVIII

	INSEM. AGE (Days)	EFA (Days)	MIN. AGE (Days)	CROWN-RUMP LENGTH (mm)	MATING	
					Single	Cont.
Stage XV						
A64-72	. . .	32	31	7.2		×
A66-75	32	30	30	7.4	×	
A68-176	. . .	32	31	8.4		×
Stage XVI						
A64-101	33	33	32	9.0	×	
A65-116	33	33	32	6.9	×	
A65-170	. . .	33	32	7.7		×
A66-87	36	34	34	8.5	×	
Stage XVII						
A64-122	36	34	33	12.7		×
A65-336	. . .	36	35	11.6		×
A68-161	36	36	35	10.8	×	
A68-252	. . .	39	39	12.2		×
Stage XVIII						
A64-95	42	36	35	16.6		×
A64-96	39	39	38	16.5		×
A65-197	. . .	41	40	14.8		×
A68-74	39	37	37	15.6	×	
A65-199	37	35	35	14.7	×	

In Stage XV embryos, the pore connecting the lens vesicle to the surface epithelium is no longer recognizable, although the lens is still attached. The nasal pits are first detectable externally, having receded as oval depressions directly laterally. In the older specimens the borders are elevated, the depth of the depression varying between specimens (figs. 7.1, 7.2). The maxillary arch is short and inconspicuous, and the dorsal and ventral segments of the mandibular and hyoid arches are detectable. The cervical sinus is deepened, and the 3d and 4th visceral arches are less conspicuous (see fig. 7.7). The arm buds are elongated. In the older specimens, they are subdivided by a slight constriction into a distal hand segment and a proximal arm-shoulder segment. The leg buds also are increasing in length and in the oldest specimens are approaching the point of exhibiting regional differentiation. The cervical flexure attains nearly a 90° angle. As this angle becomes more acute, the roof of the myelencephalon flattens (figs. 7.1 and 7.2) and does not bulge dorsally as is commonly seen in human embryos.

The nasal pit of Stage XVI embryos faces ventrally and the pit is no longer visible in profile view because of the lateral nasal process which now forms the lateral boundary. The maxillary arch is longer than in the previous stage and forms a prominent ridge on the surface when seen in profile. It blends with the dorsal

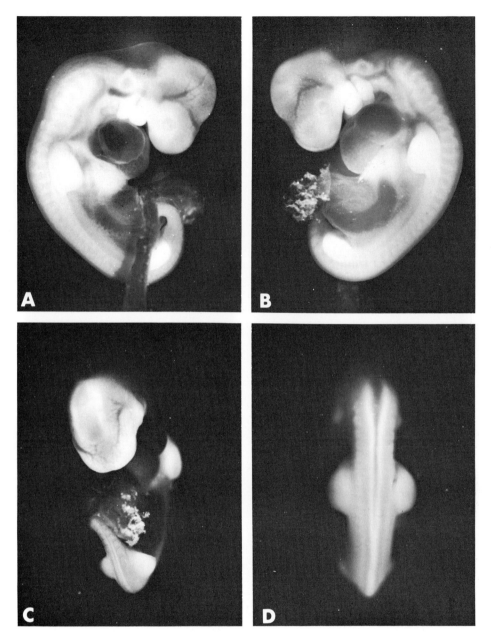

Fig. 7.1. Photographs showing the right, left, ventral, and dorsal views of the external characteristics of a younger embryo belonging to Stage XV. The primordium of the antitragus is distinguishable as a ventral segment of the right hyoid arch, and the crus and tragus are faintly visible on the right mandibular arch (A). The right atrium is devoid of blood, but the left atrium is full of blood. The liver stands out as a whitish mass just beneath the heart. The primary head veins are clearly visible in the side and ventral views. The vitelline stalk partially covers the right leg bud (A). (×10.5)

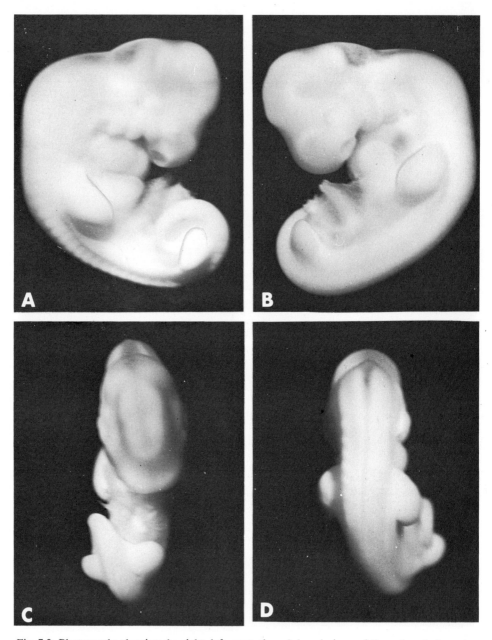

Fig. 7.2. Photographs showing the right, left, ventral, and dorsal views of the external character-istics of an older embryo belonging to Stage XV. (×7.5)

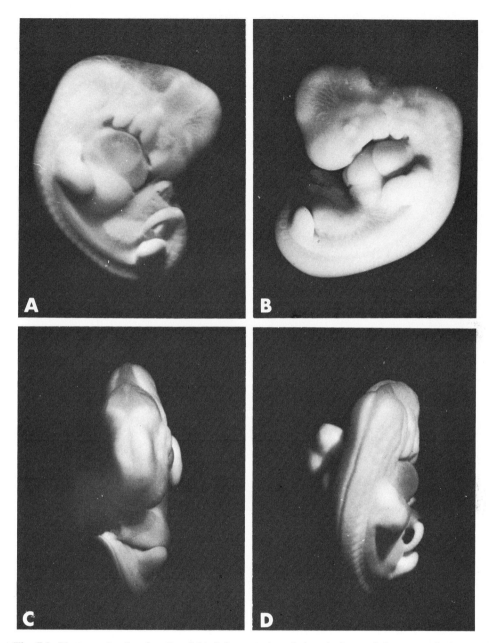

Fig. 7.3. Photographs showing the right, left, ventral, and dorsal views of the external characteristics of an embryo belonging to Stage XVI. The nasomaxillary groove and nasolacrimal
groove are clearly shown. The nasolacrimal groove, primordium of the nasolacrimal duct, extends parallel and medial to the nasomaxillary groove. (×7)

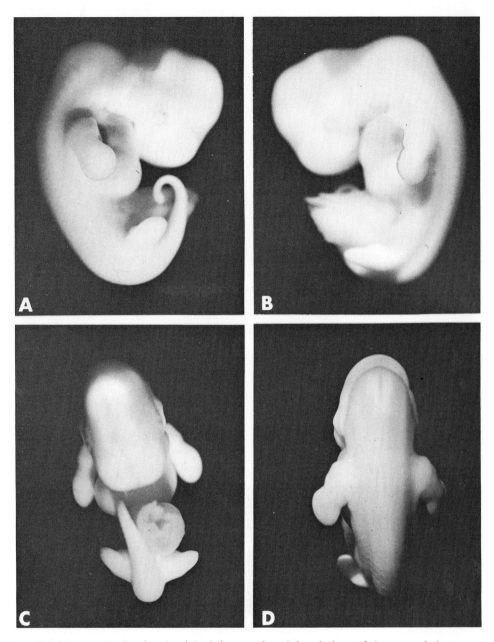

Fig. 7.4. Photographs showing the right, left, ventral, and dorsal views of the external characteristics of a younger embryo belonging to Stage XVII. (×6)

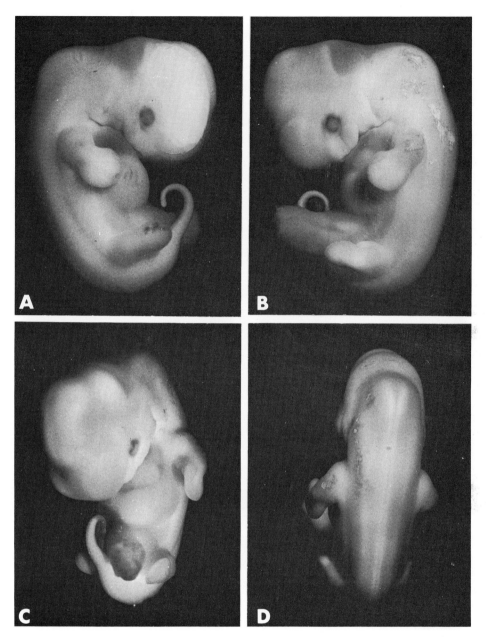

Fig. 7.5. Photographs showing the right, left, ventral, and dorsal views of the external characteristics of an older embryo belonging to Stage XVII. Note the pigmentation in the retina of the eye. (×6)

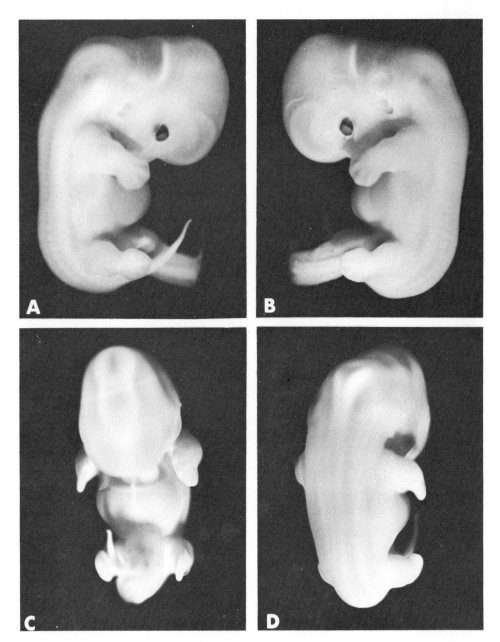

Fig. 7.6. Photographs showing the right, left, ventral, and dorsal views of the external characteristics of an embryo belonging to Stage XVIII. The auricular hillocks, which will form the pinna (*B*), and the primordium of the concha and external auditory meatus are distinct. The digital rays are visible in both the hand and foot plates, and the crenated rim of the hand plate is clearly defined. (×5)

border of the nasal pit and eye (fig. 7.3*B*). In frontal view, the upper jaw consists of a small, maxillary ridge and the primordium of the primary palate (see fig. 7.11, *XVI*). The nasofrontal groove appears as a shallow groove and marks the frontal border of the nose. The nasolacrimal and nasomaxillary grooves also are present (fig. 7.3*A*). The mandibular and hyoid arches remain prominent, but the subdivision into dorsal and ventral segments is not conspicuous. Streeter (1948) suggests that this is because the ventral ends are drawn inward and are less prominent in profile view. The expansion of the hyoid arch has caused further recession of the 3d branchial arch. The distal segment of the arm bud is beginning to form a crescentic flange, the primordium of the digital plate. The subdivision of the proximal portion of the arm bud into arm and shoulder segments is represented as a more constricted area between the digital plate and the more medial shoulder segment. The subdivision of the leg bud into a distal foot segment and a proximal leg-thigh segment, has occurred in the older specimens.

The trunk of Stage XVII embryos is straighter than that of the previous stage, but the cervical flexure is still acute. The nasal pits open towards the front, and all but the dorsal and lateral borders are shielded by the enlarging heart. In frontal view, the upper jaw and nose primordia are recognizable (see fig. 7.11, *XVII*). The maxillary arch is enlarged and extends ventromedially. The nasal pits are located more medially, and both the lateral and medial nasal processes are more obvious than in the previous stage. The auricular hillocks consist of 6 elevations, 3 on the surface of the mandibular arch and 3 on the hyoid arch. The latter are faintly recognizable in figures 7.4 and 7.5. The primordium of the external acoustic meatus appears as a shallow depression between the hyoid and mandibular elevations. In the hand, the finger rays are becoming visible and the rim of the plate shows signs of crenation. The leg bud consists of a rounded foot plate and leg and thigh regions; the first indication of a pelvic girdle appears as a condensed mass at the junction of the leg and trunk. Elevations of the individual somites are faintly visible in the lower thoracic region and are well marked from the lumbosacral region to the tip of the tail.

The trunk of Stage XVIII embryos is straighter than those of the previous stage. As a result of this, the head is beginning to move upward, decreasing the cervical angle. The retinal pigment is now conspicuous externally. In some specimens, the retina is partially overlapped by the upper and lower eyelids which are making their appearance. The nose is set off from the forehead by the frontonasal angle. In profile view, the nasomaxillary groove appears as a faint line (fig. 7.6*A*). In frontal view, it can be seen that the nostrils are closer together and separated by the medial nasal processes and nasal septum (see fig. 7.11, *XVIII*). The nasal tip or apex appears just above these structures. The upper lip is taking shape as the maxillary and premaxillary areas coalesce. The auricular hillocks are prominent elevations especially on the hyoid arch and are merging to form the auricle (fig. 7.6*B*). Distinct finger rays are present in the hand plates with interdigital notches (crenations) on their rims. The elbow is recognizable in the older specimens. Toe rays appear in the foot plate but there is no sign of crenation. The forelimbs have moved upward and partially cover the face, and the axes of both the arms and the legs are almost at right angles to the dorsal line of the body. Arms and legs are rotated in a medial direction; so the palmar surfaces face mediocaudally while the plantar surfaces face medially and slightly cranially. Somite elevations are visible in the lumbosacral region. Slight depressions mark the occurrence of intersomitic grooves through the thoracic area. The half curl of the tail, common in earlier stages, disappears and the tail remains relatively straight. It now tapers to a fine point rather than ending in a blunt knob as it did in previous stages.

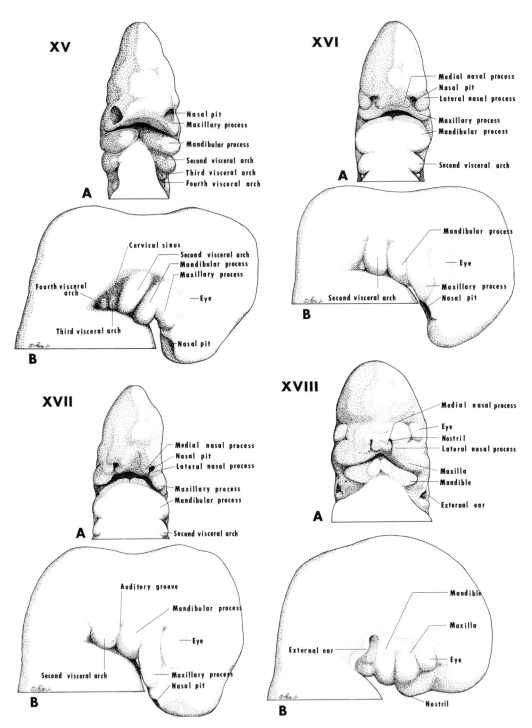

Fig. 7.7. Drawings taken from reconstructions of the face in embryos of Stages XV–XVIII. (×18) *A*, frontal views. *B*, right views.

III. Internal Characteristics

Stained, serial sections of the embryos were examined microscopically. The structures discussed below were examined for their level of development. Each specimen was subsequently placed in a particular stage based on the combined level of development of these structures.

A. *Eye (Optic Cup and Lens Vesicle)* (Fig. 7.8)

Significant changes occur in the internal structure of the eye, making it one of the better organs to use as a staging characteristic. The lens vesicle separates from the surface epithelium and its cavity disappears as the lens proper develops. Intracellular pigment granules appear in the outer, thinner pigment layer of the optic cup while the inner, thicker neural layer invaginates with the detaching lens.

In Stage XV embryos, the optic cup is separated from the brain wall by a definite optic stalk. The first indication of real pigmentation is in the form of small brownish black granules in the outer layer of the optic cup, near the rim. In this area, the outer and inner layers are in contact, decreasing the optic cavity. The rim of the optic cup extends beyond the equator of the lens, and the retinal fissure extends into the optic stalk. The lens pit is closed, and the lens vesicle is in contact with the surface ectoderm. The deep, or proximal wall of the lens vesicle is thickened to form a lens body which contains about 3 rows of elongated cells, constituting the early lens fibers. The nuclei are oval and oriented radially, and there is a nucleus free zone near the cavity of the vesicle. In the more advanced specimens, the surface ectoderm, which contains 2 to 3 rows of round nuclei, is restored where the lens vesicle has separated from it and now constitutes the epithelium of the future cornea.

The outer layer of the optic cup in Stage XVI embryos contains 3 to 4 layers of cells and more numerous pigment granules than in the previous stage. The inner layer is similar to that found in Stage XV. The lips of the retinal fissure are in close contact and in some areas are at the point of fusion. The lens vesicle is still in contact with the surface ectoderm; and the cavity of the vesicle, which is still relatively large, contains cell remnants. The lens body is thicker than that of the previous stage and contains 3 to 4 rows of elongated nuclei. The nucleus free zone, constituting the early lens fibers, is more evident.

In Stage XVII, the retinal pigmentation is more extensive and is forming near the optic stalk, the last place for it to appear. The lips of the retinal fissure are fused in the optic cup. The lens cavity is smaller and somewhat crescentic, and the lens body contains longer lens fibers. Cellular remnants still appear in the lens cavity. The lens vesicle remains in contact with the surface epithelium. The space between the rim of the optic cup, edges of the lens vesicle, and the surface ectoderm is filled by mesoderm which is continuous with that of the orbit but is less condensed.

The pigment granules, found mainly in the outer part of the retina in Stage XVIII embryos, are beginning to migrate forward. The lens fibers are distinct, and the nuclei of the lens body, comprised of about 5 rows, form a curve that is convex toward the crescentic lens cavity. The lens cavity is reduced to a slit, but its outline is visible in the embryo illustrated in figure 7.8*D*. The other embryos comprising this stage show more characteristic lens cavities.

The lens epithelium and surface ectoderm are separated. The mesenchyme between them is more condensed than in Stage XVII, forming the primordium of the cornea. The surface ectoderm is comprised of a superficial squamous layer and a cuboidal layer which lies next to the basement membrane.

XV

XVI

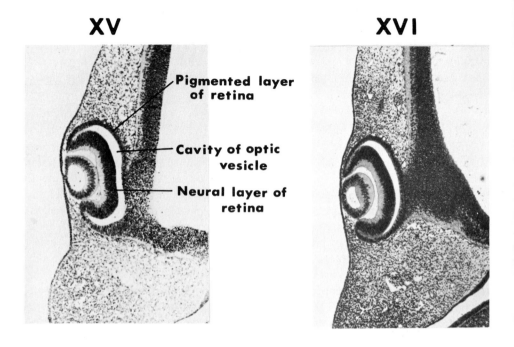

Pigmented layer of retina

Cavity of optic vesicle

Neural layer of retina

XVII

XVIII

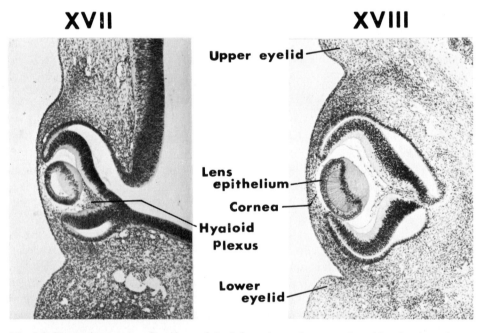

Upper eyelid

Lens epithelium

Cornea

Hyaloid Plexus

Lower eyelid

Fig. 7.8. Photomicrographs of sections of the left ocular region near the midsection in embryos belonging to Stages XIV–XVIII. The general features of the eye and related structures are shown. ($\times 55$)

B. *Inner Ear* (*Vestibular and Cochlear Pouches*) (Fig. 7.9).

The major components of the definitive internal ear differentiate in Stages XV through XVIII. The *endolymphatic duct,* the *vestibular pouch,* and the *cochlear pouch* appear by Stage XV. Further, in Stage XVI through XVIII the *semicircular ducts* and *utricle* differentiate from the *vestibular pouch* while the *saccule* and *cochlea* differentiate from the cochlear pouch.

In Stage XV, the endolymphatic duct is distinct and set off from the remainder of the otocyst. It is relatively short and joins the otocyst on its medial surface but elongates rapidly in Stages XVI through XVIII, quadrupling its length. Its walls are gradually reduced to a single cell layer distally, but the original structure is retained proximally. The primordium of the vestibular pouch appears in Stage XV as the expanded dorsal portion of the otocyst. Grooves first appear on the lateral surface of the vestibular pouch in the more advanced embryos of Stage XV, marking the areas (absorption foci) where absorption will occur in the formation of the semicircular ducts. The medial and lateral walls of the vestibular pouch are also becoming thinner. The primordium of both the superior and caudal semicircular ducts appear in Stage XVII, and 1 to 3 semicircular ducts are differentiated by Stage XVIII. The ventral portion of the otocyst, the cochlear pouch, is a slightly oblong cone in Stages XV and XVI. In Stage XVII, it extends a short distance forward and the cochlear duct resembles an L in most Stage XVIII embryos. The utricle and the saccule become apparent as the semicircular ducts and cochlear duct differentiate, respectively.

C. *Hypophysis* (Fig. 7.10)

In Stages XV through XVIII, the infundibular process of the forebrain, the anlage of the neurohypophysis or pars nervosa, develops rapidly and the hypophyseal (Rathke's) pouch, precursor of the adenohypophysis, forms a cup-shaped structure surrounding the ventral and cranial portions of the infundibulum.

1. *Adenohypophysis.* The caudal region of the hypophyseal pouch becomes flattened and spreads laterally along the wall of the forebrain in the region of the early infundibulum in Stage XV (fig. 7.10*A, B*). As the infundibulum grows into a fingerlike projection during Stages XV to XVIII, the adenohypophysis surrounds the rostroventral portion forming a double layered, cuplike structure, similar to the optic cup. The caudal region in contact with the infundibulum is farthest removed from the connection with the pharynx and forms the early pars intermedia of the adenohypophysis. Although the wall is thickened, compared to the pharyngeal epithelium, the thickening is much less pronounced than in the more rostral regions. The thick rostral wall forms the primordium of the pars distalis (fig. 7.10*C*), but no cell cords or trabeculae can be found in any of the embryos of Stage XVIII. The pars tuberalis is first recognizable at Stage XVII as short rostrolateral extensions of the adenohypophyseal wall in the area where the pouch remains open to the pharyngeal cavity (fig. 7.10*D*). The common lumen between the adenohypophyseal and pharyngeal cavities is closed by Stage XVIII although many of the embryos still have solid tissue connections with the pharyngeal epithelium.

2. *Neurohypophysis.* The infundibulum is present at Stage XV, somewhat earlier than found in man, as a relatively thin-walled depression in the floor of the forebrain. This depression rapidly elongates into a distinct, fingerlike, ventral projection and maintains a lumen common with the forebrain through Stage XVIII (fig. 7.10*B*). In some embryos of Stage XVIII, the thickened wall begins to fold.

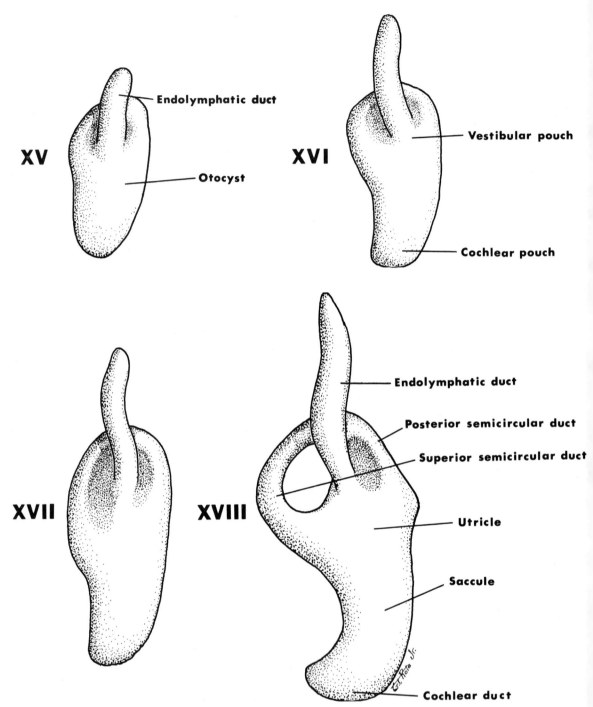

Fig. 7.9. Drawings taken from reconstructions of the left otocyst in embryos of Stages XV–XVIII. (×80)

116

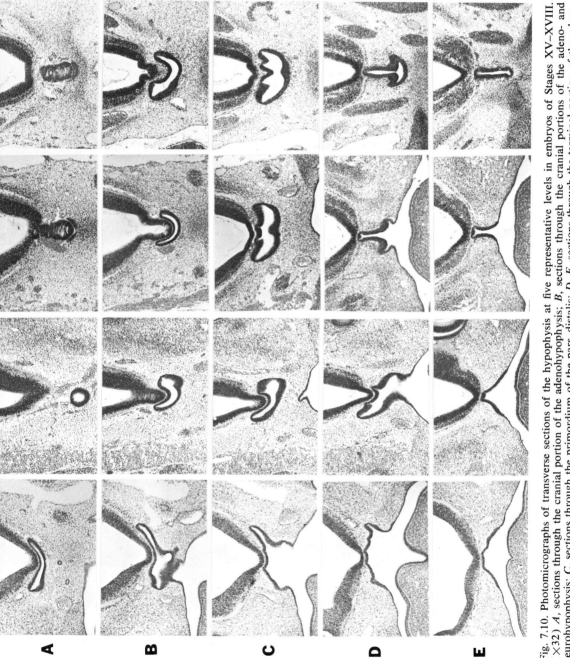

Fig. 7.10. Photomicrographs of transverse sections of the hypophysis at five representative levels in embryos of Stages XV–XVIII. (×32) A, sections through the cranial portion of the adenohypophysis; B, sections through the cranial portions of the adeno- and neurohypophysis; C, sections through the cranial portions of the adeno- and neurohypophysis; D, E, sections through the terminal portion of the adeno- and neurohypophysis.

D. *Palate and Adjacent Areas* (Fig. 7.11)

The nasal pits and the medial and lateral nasal processes become established early, followed by the appearance of the maxillary processes and the nasal fin. Nasal fin degeneration begins in Stage XVII as the primary palate forms. The primordium of the nasal septum appears with the bucconasal membrane.

The nasal placodes have become the nasal pits by Stage XV because of proliferation of the mesenchyme that surrounds them. They are widely separated on the frontal surface of the head and are bordered by the medial and lateral nasal processes (fig. 7.11, *XV*). In Stage XVI embryos, the maxillary process has developed cranially and contributes to the ventrolateral margin of the nasal pit, which continues to sink beneath the surface (fig. 7.11, *XVI*). The epithelium separating the mesenchyme of the maxillary and medial nasal processes forms the nasal fin which is continuous, cranially, with the epithelium lining the nasal pit and, caudally, with the epithelium lining the roof of the developing oral cavity.

During Stage XVII, the nasal fin expands. Cranially, it includes the epithelia of the lateral and medial nasal processes, and caudally, it is formed by the epithelia of the maxillary and medial nasal processes. Nasal fin degeneration occurs as the primary palate forms (fig. 7.11, *XVIII-B*). Olfactory nerve fibers appear along the craniomedial aspect of the primary nasal cavity and pass toward the telencephalon (fig. 7.11, *XVII-A*).

By Stage XVIII, the primary palate has formed as a continuous bridge of mesenchyme extending laterally from the midline area that is caudal to the developing nasal septum to the maxillary process mesenchyme (fig. 7.11, *XVIII-A*). The primary palate extends through the nasal fin epithelium, which is still present at the caudal edge of the primary palate as the bucconasal membrane (fig. 7.11, *XVIII-B*). This membrane separates the blind end of the primary nasal cavity from the bucconasal cavity. The primordium of the nasal septum cartilage appears as a condensation of mesenchymal cells in the area between the primary nasal cavities (fig. 7.11, *XVIII-A*). No vomeronasal organ is noted, but a distinct bundle of nerve fibers is observed cranial to the olfactory nerve in the area which will become the vomeronasal organ (fig. 7.11, *XVIII-A*). The epithelial remnant which will give rise to the nasolacrimal duct extends ventromedially and cranially from the nasolacrimal groove on the lateral surface of the face and ends as a solid core of epithelial cells approximately midway between the floor of the primary nasal cavity and the lateral surface of the head.

E. *Endodermal Derivatives*

1. *Trachea* (fig. 7.12). The trachea is separated from the esophagus over most of its length. The thickness of the tracheal epithelium does not increase noticeably during this period, but the total diameter increases as the lumen expands. The periepithelial wall of the trachea is much more dense than the nearby esophagus, but no differentiation into cartilage and connective tissue elements is noticed by Stage XVIII.

2. *Lung* (fig. 7.13). The lungs develop from outgrowths of the terminal bifurcations of the trachea, the primary bronchi, by Stage XV but show no signs of branching. By Stage XVI small buds appear on the lateral walls of the primary bronchi, two on the right and one on the left, which along with distal continuations of the bronchi, represent the early primordia of the endodermal portions of the three right and two left lobes of the lungs. Elongation and branching of the primary bronchi continues through Stage XVII. By Stage XVIII the cranial, middle, caudal, and cardiac bronchi are detectable.

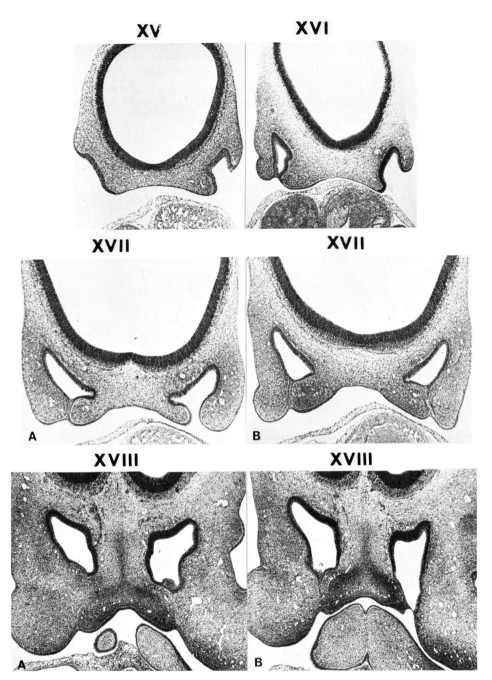

Fig. 7.11. Photomicrographs of coronal sections through various levels of the palate region in embryos belonging to Stages XV, XVI, XVII, and XVIII. *XV,* section through the olfactory pits and lateral and medial nasal processes in a Stage XV embryo; *XVI,* section through the maxillary process in a Stage XVI embryo; *XVII-A,* section through the rostral end of the lateral and medial nasal processes, showing the nasal fin in a Stage XVII embryo; *XVII-B,* section from the same embryo through the dorsal portion of the process where the nasal fin is degenerating; *XVIII-A,* section through the rostral end of the primary palate and developing nasal septum in a Stage XVIII embryo; *XVIII-B,* section from the same embryo through the dorsal portion of the primary palate where the bucconasal membrane has formed. ($\times$30)

Dense fields of mesenchyme surround the bronchi and developing lungs throughout the period from Stages XV to XVIII. Clusters of angiogenic cells and some open capillaries are present within this tissue at Stage XV, and by Stage XVIII the area is highly vascular. Coalescence of these capillaries forms a large common pulmonary vein which can be traced into the left atrium of the heart.

3. *Esophagus* (fig. 7.12). The esophagus is elongated and narrowed from earlier stages. The epithelium almost doubles in thickness from Stage XV to Stage XVIII and the esophageal lumen increases in diameter. At Stage XV, a zone of

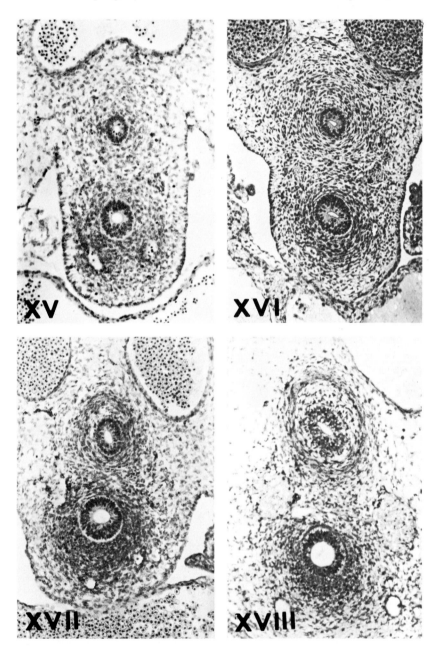

Fig. 7.12. Photomicrographs of transverse sections taken at comparable levels through the trachea and esophagus at Stages XV–XVIII. Note the histogenesis in the zones surrounding both tubes. The paired vagus nerves are evident lateral to the junction of the esophagus and trachea, and the dorsal aortae and pulmonary arteries are prominent above and below both structures respectively. (×120)

dense circumferentially arranged mesenchyme is evident around the epithelium. This zone becomes progressively more distinct and by Stage XVIII is subdivided into a wide, less dense, submucous zone, which lies against the epithelium, and an outer dense muscular coat.

4. *Pancreas* (fig. 7.14). The pancreas develops from two separate evaginations of the foregut which eventually fuse into a single organ. The dorsal pancreatic diverticulum is first recognizable in Stage XII as a thickened area of the gut epithelium, dorsal to the liver diverticulum. By Stage XIII, there is a distinct outpocketing of the gut wall in this area.

The dorsal pancreatic diverticulum is slightly constricted at its junction with the intestinal wall by Stage XIV (fig. 7.14*A*) and the ventral pancreatic diverticulum becomes recognizable as a small evagination from the caudal surface of the common bile duct. The dorsal pancreatic diverticulum elongates rapidly and to a much greater extent than the ventral pancreatic diverticulum. By Stage XVI the dorsal pancreatic diverticulum is a prominent mass extending craniolaterally from the left side of the duodenum (fig. 7.14*B*). It retains its connection to the duodenum by a narrow duct at a point distinctly caudal to the insertion of the common bile duct. The ventral pancreatic diverticulum remains much smaller than the dorsal pancreatic diverticulum and connects with the common bile duct near the latter's insertion into the duodenum. Rotation of the entire biliary duct system is evident so that the ventral pancreatic diverticulum and the portion of the common bile

XVI   **XVIII**

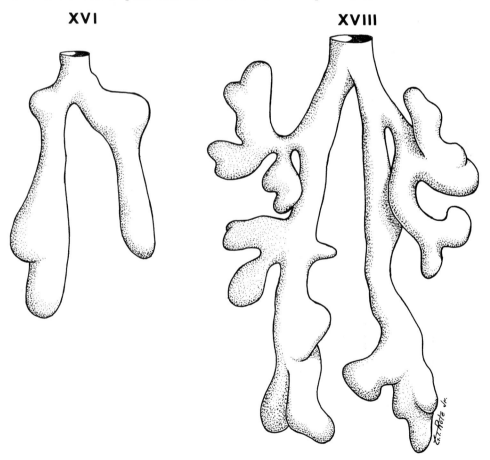

Fig. 7.13. Drawings taken from reconstructions of the lungs in embryos belonging to Stages XVI and XVIII. Ventral view. (×60)

duct proximal to the duodenum are shifted to a position dorsolateral to the right side of the duodenum.

The continued dorsal rotation of the common bile duct and ventral pancreatic diverticulum dominates the further changes in the pancreas. By Stage XVII, the ventral pancreatic diverticulum has reached a position directly dorsal to the duodenum (fig. 7.14C). In this position it has come into contact and fused with a small proximal portion of the dorsal pancreatic diverticulum. The ventral pancreatic duct joins the common bile duct near this point of fusion, but the duct of the dorsal pancreas continues separately to join the duodenum caudal to the insertion of the common bile duct.

The ventral pancreas of the baboon remains very small compared with man, and the degree of fusion with the dorsal pancreas is considerably reduced. After fusion of the two pancreatic diverticuli, portions of the dorsal pancreas are associated with the duct of the ventral pancreas, which becomes the main pancreatic duct (*ductus pancreaticus* or duct of Wirsung). The dorsal pancreatic duct (*ductus pancreaticus accessorius* or duct of Santorini), however, is retained in the adult (Boyden 1967) rather than disappearing before birth as in man. In addition, it enters the duodenum caudal to the insertion of the common bile duct, in contrast to the cranial insertion found in man.

5. *Caecum* (fig. 7.15). The caecum is a useful criterion by which to classify specimens for Stages XV to XVIII. However, a vermiform appendix in the form of a blind tubular sac as appears in human embryos was not observed and, con-

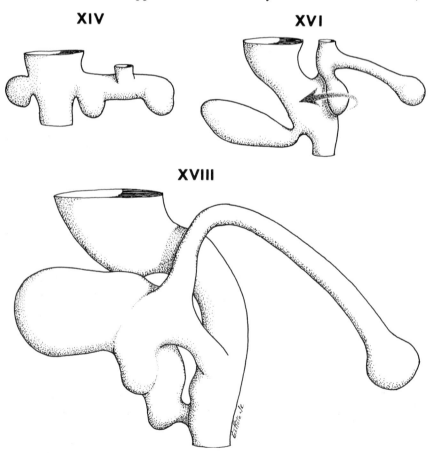

Fig. 7.14. Drawings taken from reconstructions of the developing pancreas in embryos of Stages XIV, XVI, and XVIII. (×60)

sequently, is not a usable criterion for staging baboon embryos. At Stage XV, the caecal swelling appears as a small conical dilation of the ascending or caudal portion of the primary intestinal loop. Its appearance marks the junction of the ileum and ascending colon, which at this stage is located in the umbilical cord. By Stage XVI, the caecum has expanded in diameter. In Stage XVII embryos the primary intestinal loop projects farther into the umbilical cord and the caecum appears as a diverticulum (fig. 7.15*B*). By Stage XVIII, the caecum has elongated and projects ventrocaudally (fig. 7.15*C*). The vitelline duct communicates with the intestine at the most ventral extension of the primary intestinal loop in Stage XV, but by Stage XVI it has atrophied.

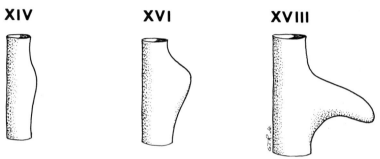

XIV   **XVI**   **XVIII**

Fig. 7.15. Drawings taken from reconstructions of the caecum in embryos of Stages XIV, XVI, and XVIII. (×60)

F. *Metanephros* (Fig. 7.16)

Previous to Stage XV the metanephric diverticulum arises from the mesonephric (Wolffian) duct near its cloacal end. It grows dorsally and cranially into the caudalmost portion of the nephrogenic mesoderm where the metanephrogenic mass forms.

By Stage XV, the ureter has elongated and expanded at its distal end. Mesenchymal cells comprising the metanephrogenic mass condense around its expanded distal end (fig. 7.16, *XV*). The elongating ureter is expanding into a renal pelvis in Stage XVI (fig. 7.16, *XVI*). In the older specimens, upper and lower poles are differentiating and the metanephrogenic mass is condensed and thickened. Calyces appear in Stage XVII as the renal pelvis expands and divides, pushing into the metanephrogenic mass which has not subdivided (fig. 7.16, *XVII*). The kidney in Stage XVIII embryos is enclosed in a renal capsule primordium and the ureter is elongated. Multiple calyces branch from the renal pelvis, and collecting tubule primordia appear (fig. 7.16, *XVIII*). The metanephrogenic mass subdivides and a small mass, the nephron primordium, becomes associated with each terminal collecting tubule. Although the developing kidney of the baboon equals that of man stage by stage, its characteristics seem to be more similar to those of the younger specimens and never quite attain the degree of development of the older specimens for a given stage.

G. *Müllerian Duct*

The Müllerian, or paramesonephric, duct is identifiable in Stage XVI embryos as a thickening of the coelomic epithelium lateral to the Wolffian or mesonephric duct. The primordium of the tubal ostium appears slightly cranial to the mesonephric duct by Stage XVII, and evagination of the thickened coelomic epithelium begins. By Stage XVIII, the duct ranges from 0.2 to 0.7 mm in length. Its usefulness in classifying embryos is limited to Stage XVIII embryos (Streeter 1948).

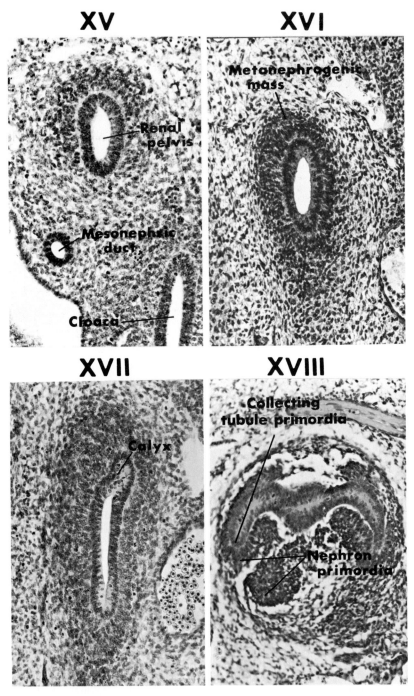

Fig. 7.16. Photomicrographs of sections of the metanephros in embryos of Stages XV–XVIII. ($\times 150$)

References

Boyden, E. A. 1967. The choledochoduodenal junction in the Kenya Baboon (*Papio cynocephalus* and *Papio anubis*). In *The Baboon in Biomedical Research,* ed. H. Vagtborg, 2:117–32. Austin: University of Texas Press.

Streeter, G. L. 1948. Developmental horizons in human embryo: Description of age groups XV, XVI, XVII, and XVIII. *Contrib. Embryol., Carneg. Inst.* 32:135–203.

8

Description of Stages XIX, XX, XXI, XXII, and XXIII

Raymond F. Gasser/Andrew G. Hendrickx/Joe A. Bollert

I. Age and Size

STAGE XIX—EFA, 39 ± 1 days; CR length, 16–17 mm.
STAGE XX—EFA, 41 ± 1 days; CR length, 17–18 mm.
STAGE XXI—EFA, 43 ± 1 days; CR length, 18–21 mm.
STAGE XXII—EFA, 45 ± 1 days; CR length, 21–23 mm.
STAGE XXIII—EFA, 47 ± 1 days; CR length, 25–28 mm.

Data on the age and size of the embryos of Stage XIX through XXIII are given in table 8.1. The embryos are listed in the order of their morphological development. There is a close correlation between the estimated fertilization age and morphological development in all but 4 specimens. Of these 4 specimens, A65-103 and A65-97 (XXI) are 1 day older, A67-265 (XXIII) is 1 day younger, and A64-94 (XXIII) is 2 days older than the estimated fertilization age of each respective stage. Age and size are well correlated with only 2 inconsistencies. A68-160 (XXI) and A68-9 (XXIII) are slightly larger and are less advanced than the other specimens in their respective stages. This is partly explained by the use of different fixatives. The variation in crown-rump length in Stages XIX through XXIII is 0.5, 1.0, 3.7, 0.8, and 3.0 mm, respectively. It is consistent with the preceding stages and is within the range reported for human embryos (Streeter 1951). Of the 9 single matings, 6 occurred within 1 day of optimal mating time (the 3d day preceding deturgescence) and 3 occurred on the 5th day preceding deturgescence. All 3 of the embryos derived from mating on the 5th day preceding deturgescence represent the younger embryos in their respective stages. In all instances the estimated fertilization age corresponds more closely with the developmental characteristics than does the insemination age.

II. External Characteristics

Many significant features develop externally, some of which give the specimens a distinct subhuman form (figs. 8.1 to 8.5). In the head region a snout with nares becomes evident, the eyelids and auricles develop, and the rima oris is well defined. The neck appears as the cervical angle is reduced. The digits differentiate on the fore- and hindlimbs, and both appendages undergo rotation and subdivide. The tail becomes more distinct.

In Stage XIX the upper jaw protrudes beyond the plane of the forehead

TABLE 8.1

EMBRYOS OF STAGES XIX–XXIII

	INSEM. AGE (Days)	EFA (Days)	MIN. AGE (Days)	CROWN-RUMP LENGTH (mm)*	MATING	
					Single	Cont.
Stage XIX						
A65-107	40	38	38	16.8	×	
A64-91	40	40	39	16.3	×	
Stage XX						
A67-271	43	42	41	17.2		×
A68-93	...	40	39	17.7		×
A68-5	...	41	40	18.2		×
Stage XXI						
A67-281	43	43	42	19.6	×	
A68-160	44	42	41	21.7	×	
A67-286	44	42	41	18.0	×	
A65-103	45	45	44	18.3	×	
A65-97	...	45	44	19.4		×
Stage XXII						
A67-31	46	46	45	21.7	×	
A68-53	...	47	46	22.5		×
A68-158	...	47	46	22.3		×
A68-4	...	47	46	22.0		×
Stage XXIII						
A67-265	45	45	45	25.0	×	
A68-9	...	46	45	28.0		×
A65-127	49	47	46	27.6		×
A68-97	...	47	46	26.8		×
A64-94	50	50	50	28.0	×	

* Taken after fixation.

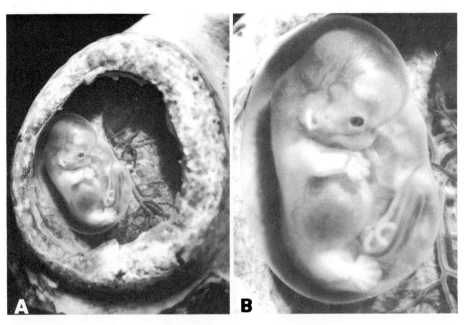

Fig. 8.1. Photographs showing the right views of the external characteristics of an embryo belonging to Stage XIX. *A, in situ,* note the amnion, vitelline sac, and cut edge of chorion. (*A,* ×1.8; *B,* ×4.5)

(fig. 8.1). The protrusion of the lower jaw does not parallel that of the upper jaw but is more advanced than in the previous stage. The median notch of the mandibular arch is no longer evident. The upper and lower eyelids are narrow folds at the margins of the eye. The auricular hillocks are coalescing and are less distinct. Epithelial plugs appear in the external nares that are well formed and open laterally. As the back straightens the cervical angle becomes less acute. Constrictions at the elbow and wrist divide the forelimb into arm, forearm, and hand. There is some bending at the elbow. The palmar surface of the hand points caudomedially. Interdigital notches are distinct on the hand but are just beginning on the foot.

The face of Stage XX embryos has a smoother contour than the specimens.

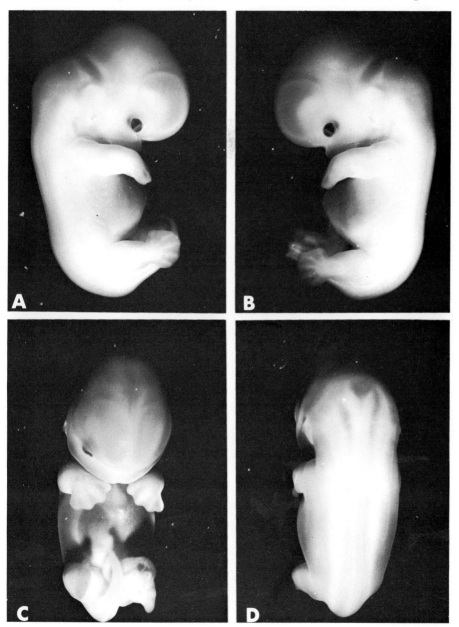

Fig. 8.2. Photographs showing the right, left, ventral, and dorsal views of the external characteristics of an embryo belonging to Stage XX. ($\times 4.5$)

of the previous stage because of blending of the maxilla with the premaxilla (fig. 8.2). The auricular hillocks have joined into one mass which surrounds the external auditory meatus and contains cartilage. The trunk lengthens and straightens slightly with a lifting of the head. The fore- and hindlimbs lengthen with the bends at the wrist and elbow becoming more pronounced. The hands are more medial in position, located just lateral to the nose and ventral to the heart. The palms are directed more caudally. The interdigital notches of the hand have deepened and the short finger is spread laterally. For the most part, the position of the hindlimb is unchanged. The interdigital notches of the foot are present, and both the thumb and great toe are distinguishable as apposable digits.

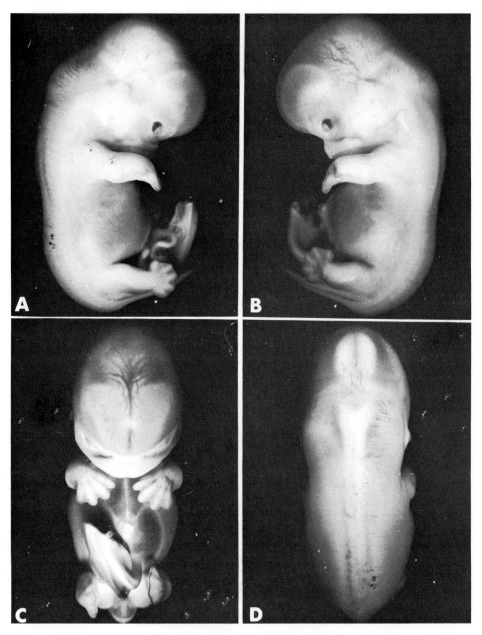

Fig. 8.3. Photographs showing the right, left, ventral, and dorsal views of the external characteristics of an embryo belonging to Stage XXI. ($\times 3.5$)

The face continues to lengthen in Stage XXI (fig. 8.3). The upper and lower jaws elongate at the same rate, but the upper jaw remains further extended. With decrease of the interdigital tissue the fingers become longer and closer together. The toes become more evident as the interdigital notches of the foot deepen. The vascular plexus at the periphery of the hand and foot is obvious.

In Stage XXII jaw protrusion is prominent with the lower jaw growing more rapidly than the upper jaw (fig. 8.4). The eyelids thicken, and the tragus and antitragus are apparent in a more definitive auricle. With additional lengthening of the limbs the fingers on one side overlap those on the other side, and the feet

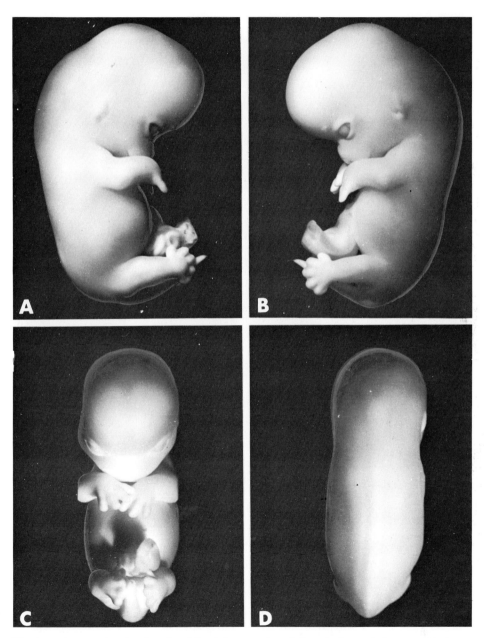

Fig. 8.4. Photographs showing the right, left, ventral, and dorsal views of the external characteristics of an embryo belonging to Stage XXII. (×3.5)

move closer together. Touch pads appear as slightly raised areas on the terminal phalanges of the fingers.

With extension of the neck the head is raising away from the ventral aspect of the chest in Stage XXIII (fig. 8.5). More definite contours appear in the face. The lower jaw protrudes almost as far as the upper jaw. The eyelids cover most of the eye, and the auricle is assuming its definitive shape. The length of the extremities increases considerably. The forearm is raised along with the head so that it is above the shoulder. The hands overlap in front of the snout with the palms pointed mostly caudally. The feet approximate one another and are separated by a long tapering tail that reaches the umbilical cord. Sometimes the feet overlap. The plantar surface of the foot begins to turn caudally.

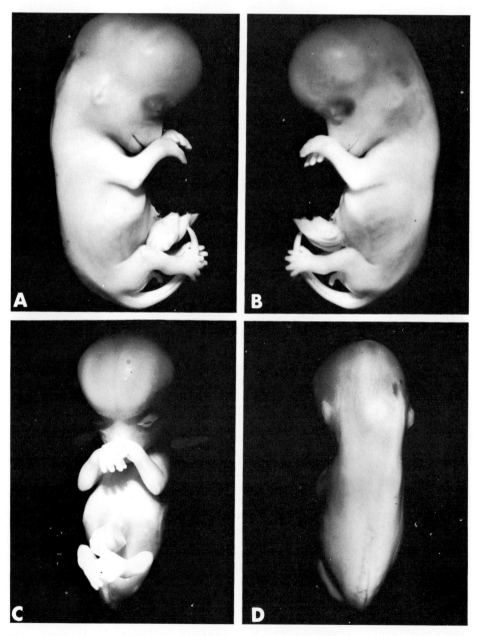

Fig. 8.5. Photographs showing the right, left, ventral, and dorsal views of the external characteristics of an embryo belonging to Stage XXIII. (×3)

III. Internal Characteristics

Stained, serial sections of the embryos were examined microscopically. The structures discussed below were examined for their level of development. Each specimen was placed subsequently in a particular stage based on the combined level of development of these structures.

A. *Eye*

   The eye progressively increases in size. The inner and outer layers of the optic cup show significant differentiation. Both layers approximate one another but remain apart in some areas by the primitive cavity of the optic vesicle (intra-retinal space). The vascular (choroidea) and fibrous (sclera) coats form around the outer layers. The optic stalk transforms into the optic nerve, and three layers of the cornea are established. Accessory ocular structures such as the eyelids and lacrimal glands appear, and most of the definitive nerves of the orbit, the ciliary ganglion, and the extrinsic ocular muscles can be identified.

   1. *Optic cup* (fig. 8.6). In Stage XIX the inner layer of the optic cup is evident as the thicker neural layer, and the outer layer is the thinner pigment layer. The neural layer can be further subdivided into an outer, nucleated (ependymal and mantle) layer and an inner, clear (marginal) layer. A vascular canal has replaced the optic fissure and the hyaloid plexus is present in the vitreous body. Lens fibers have obliterated the cavity of the lens vesicle. Vascular spaces at the periphery of the pigment layer will become much of the choroidea. The scleral condensation is forming at the periphery of the optic cup and is most prominent and densest at the corneal margin of the cup. Each eyelid is a blunt fold of tissue above or below the eye and covers approximately one-sixth of the exposed surface of the eye.

   A plexus of vessels located at the posterior aspect of the lens forms a vascular tunic in the embryos of Stage XX. The scleral tissue becomes denser and spreads more evenly around the optic cup. The eyelids are less blunt folds of tissue, each covering approximately one-fifth of the exposed eye surface.

   The anterior chamber is present between the cornea and the thin but distinct pupillary membrane in Stage XXI embryos. The pupillary membrane is adjacent to the permanent lens epithelium and joins peripherally with mesenchymal tissue that is continuous with the scleral condensation. Each eyelid grows to cover approximately one-fourth of the exposed part of the eye. In Stage XXII the sclera is a discrete layer deep to which are many blood vessels of the choroidea. Each eyelid covers as much as one-third of the exposed surface of the eye.

   The outer, nucleated layer of the neural retina begins to subdivide by Stage XXIII into a more compact outer neuroblastic layer and a less compact inner neuroblastic layer (see fig. 8.7, *XXIII-A*). Since the sclera is becoming a dense fibrous layer while the choroidea is becoming loose, spongy, and vascular, the boundaries of these two layers are more distinct. The sclera is continuous with the substantia propria of the cornea. The eyelids almost completely cover the eye and begin to fuse with each other at the angles of the eye.

   2. *Optic stalk or nerve* (fig. 8.7). A canal is present in the proximal and distal portions of the optic stalk in Stage XIX but is slightly obscure in the middle part. It enlarges as it approaches the diencephalon. The ependymal arrangement of nuclei is retained, for the most part, around the canal. The stalk is small and slender and a shallow groove on its distal segment is all that remains of the optic fissure. Nerve fibers are more evident in the stalk distally than proximally. Few, if any,

of them appear to arrive at the brain. The hyaloid vessels enter the most distal portion of the stalk where the stalk and the retina are continuous.

By Stage XX the optic canal is indistinct except for a short segment near the diencephalon where it is continuous with the 3d brain ventricle. The ependymal arrangement of nuclei is breaking up with scattering of previously compact nuclei. The groovelike remnant of the optic fissure has disappeared. Additional nerve

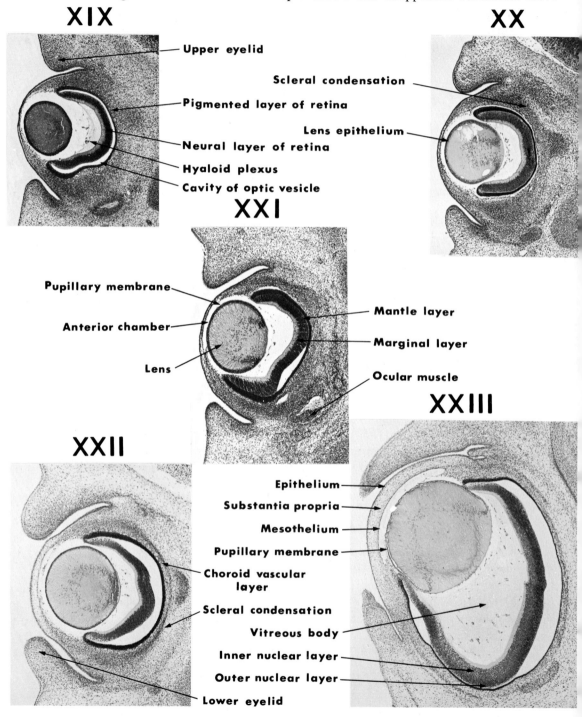

Fig. 8.6. Photomicrographs of sections of the left ocular region near the midsection of the eye in embryos belonging to Stages XIX to XXIII. The general features of the eye and related structures are shown. (×30)

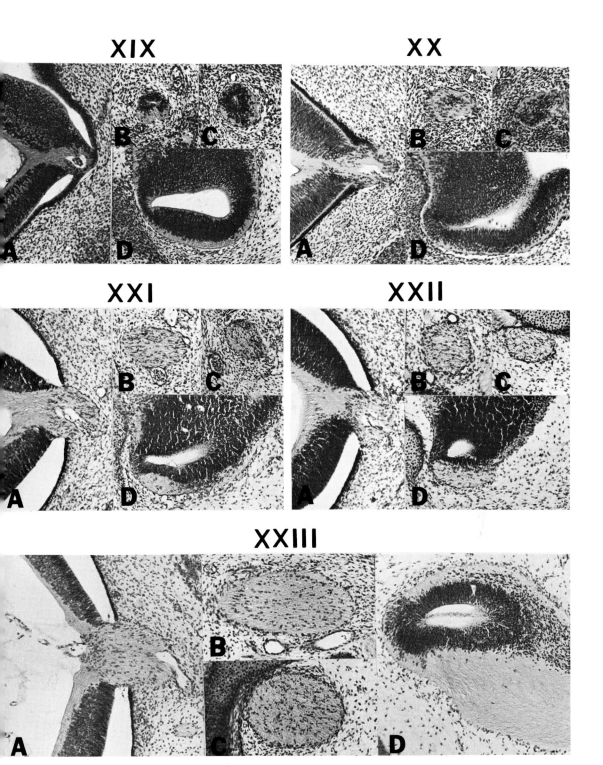

Fig. 8.7. Photomicrographs of sections through four different regions of the left optic nerve in embryos belonging to Stages XIX–XXIII. (×75) Progressive morphological changes in the nerve are shown with an indication of nerve fiber growth from the retina to the diencephalon. *A*, sections through the retina–optic nerve junction where the hyaloid vessels enter the nerve; *B*, sections through the distal portion of the nerve; *C*, sections through the proximal portion of the nerve; *D*, junction of optic nerve and diencephalon showing the most peripheral extent of the 3d brain ventricle and the formation of the optic chiasma.

fibers collect at the periphery of the stalk and some of them appear to arrive at the diencephalon. There· is a suggestion of the chiasma forming on the ventral surface of the diencephalon in the midline. Hyaloid vessels enter the stalk a short distance from the retina, and a sheath is beginning to form around the stalk.

The optic canal is a small extension of the 3d ventricle in Stage XXI embryos. The ependymal arrangement of nuclei has mostly disappeared but occasionally there are small concentrations of nuclei, mainly in the proximal segment. Nerve fibers are throughout the distal segment of what can now be called a nerve since many of the fibers arrive at the diencephalon. The chiasma is a more definite structure and a sheath is evident around distal parts of the optic nerve. Hyaloid vessels traverse a short portion of the nerve distally before reaching the retina.

At Stage XXII the extension of the 3d ventricle into the nerve has almost disappeared. Collections of nuclei within the nerve are fewer and smaller and nerve fibers course throughout. The chiasma is very evident and a sheath surrounds the entire nerve.

A very prominent optic nerve is present at Stage XXIII and appears to contain a much greater quantity of nerve fibers. Between groups of nerve fibers there are nuclei in a striate arrangement that is especially evident in the longitudinal sections of the nerve. The nerve sheath is prominent and the chiasma is large. The 3d ventricle has no real extension into the nerve. The hyaloid vessels course through a longer portion of the nerve distally.

3. *Cornea* (figs. 8.6, 8.8). The cornea begins its differentiation in Stage XIX as a thin, loose layer of mesodermal cells deep to the surface epithelium. This mesodermal layer is 1–2 cells thick at the midpoint of the eye and is continuous peripherally with the scleral condensation. The mesodermal cells become organized into a more discrete layer during Stage XX, but there is no real increase in thickness. An occasional nucleus between the mesodermal layer and the surface epithelium is the first appearance of the substantia propria.

Three corneal layers can be easily identified in the embryos of Stage XXI. The mesothelial layer (Descemet's) is 1–2 cells thick and lines part of the anterior chamber. The substantia propria contains 2–6 layers of loosely arranged nuclei and remains in continuity with the scleral condensation. The epithelial layer is well defined. At Stage XXII the mesothelial layer contains many large ellipsoidal-shaped nuclei. The substantia propria increases in thickness and is loosely arranged having large nuclei that are either spherical or ellipsoidal in shape. A thick substantia propria separates the well-defined epithelial and mesothelial layers in Stage XXIII specimens. It more than doubles in thickness and is composed of cells whose nuclear shapes range from ellipsoidal to thinly fusiform. Fibrous material is present between the cells.

B. *Cochlea*

The slender, ventral portion of the otic vesicle differentiates into the cochlear duct which is L-shaped just prior to Stage XIX. The duct subsequently changes rapidly, spirals progressively, and approaches $2\frac{1}{2}$ turns by Stage XXIII. The extent of spiraling provides another basis for staging embryos. The cochlear duct that is most representative of each stage was reconstructed at 100 diameters. The reconstructions were cast in plaster from wax molds and are illustrated in figure 8.9.

The previously L-shaped cochlear duct has a short upturned tip at Stage XIX giving it a J-shape. By Stage XX the duct is growing upward and horizontally with its tip somewhere between turning up and turning down. After additional lengthening in Stage XXI the duct tip is definitely turned down and may begin

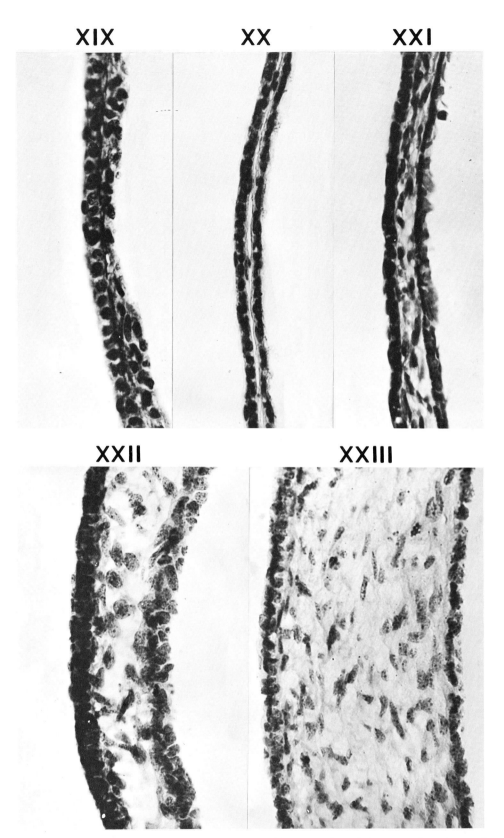

Fig. 8.8. Photomicrographs of sections through the thinnest portion of the cornea in embryos of Stages XIX–XXIII. (×465) The outer, epithelial layer is on the left, the inner, mesothelial layer is on the right, and the substantia propria is forming in between. At Stage XX there is some apparent contraction of the epithelium and mesothelium as these two layers separate slightly. The substantia propria first appears as occasional nuclei between the two layers.

its second horizontal growth. The second horizontal growth is completed by Stage XXII, and the tip of the duct is about to turn up or is definitely turned up for the second time. In Stage XXIII the duct has grown horizontally for the third time and may have a tip that is turned down for the second time. A shallow blind process is forming at the basal end of the cochlear duct, and the ductus reuniens becomes evident since the connection between the cochlear duct and the sacculus is narrowing.

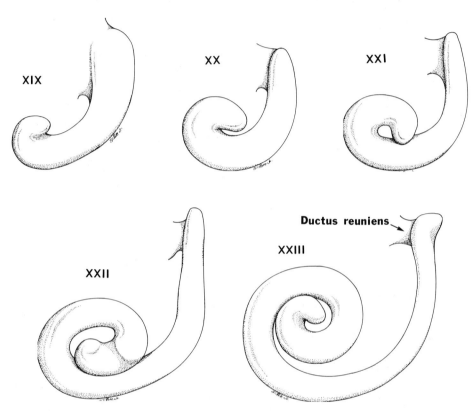

Fig. 8.9. Drawings taken from reconstructions of the left cochlear duct in embryos of Stages XIX–XXIII. (×35)

C. *Hypophysis*

Profound changes occur in the hypophyseal (Rathke's) pouch. The distal portion of the pouch develops close to and around the infundibular outpouching of the forebrain (neurohypophysis or pars nervosa) and becomes the adenohypophysis. Between cords of cells that grow from the wall of the adenohypophysis are vascular spaces and angiogenic tissue. The cords and the vascularization progressively increase in quantity and complexity (fig. 8.11). A pharyngeal hypophysis (pars pharyngea) occasionally develops from the proximal part of the hypophyseal

Fig. 8.10. Photomicrographs of transverse sections of the hypophysis at five representative levels in embryos of Stages XIX–XXIII. (×30) *A*, sections through the cranial portion of the pars nervosa and adjacent adenohypophysis; *B*, sections through the pars intermedia and nervosa; *C*, sections through the middle portion of the pars distalis; *D*, sections through the caudal portion of the pars distalis; *E*, sections through the terminal portion of the pars distalis. The changing pattern of the various parts of the gland can be appreciated by scanning the horizontal rows *A* to *E*.

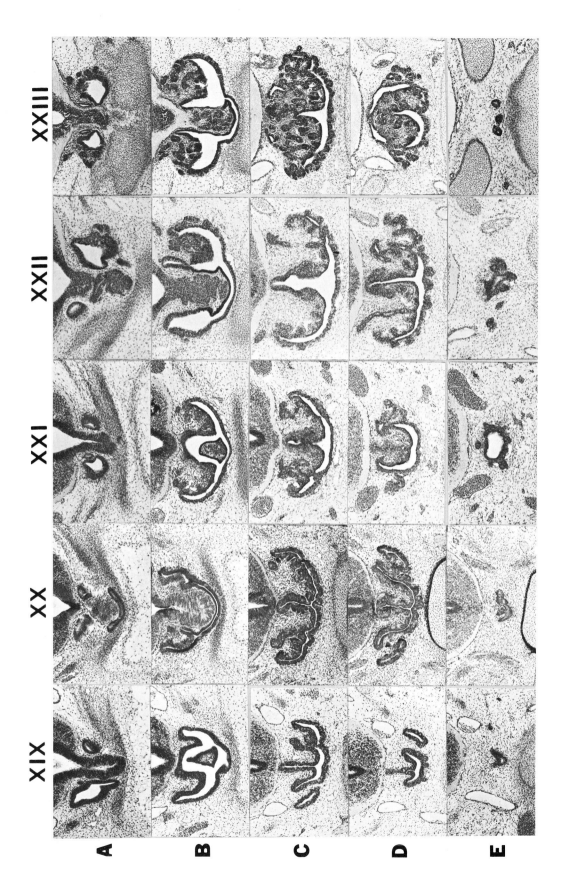

XIX   XX   XXI   XXII   XXIII

A

B

C

D

E

pouch that is attached to the roof of the pharynx (fig. 8.12). The intermediate portion of the pouch undergoes involution, for the most part, but during these stages various stem remnants are present (fig. 8.13).

1. *Adeno- and neurohypophysis.* (figs. 8.10, 8.11). The pars distalis, intermedia, and tuberalis of the adenohypophysis are distinguishable as subdivisions of the wall of the hypophyseal pouch at Stage XIX; and the pars nervosa is an outpouching of the forebrain that contains a small lumen. During subsequent stages the pars distalis becomes the most extensive portion with cords of cells, or trabeculae, growing rostrally and cranially from the original rostral wall of the pouch (fig. 8.10C, D). Some of the cords contain extensions of the original lumen, others do not. Even though the size and extent of the adenohypophysis progressively increase from stage to stage, the size of the residual lumen does not always decrease proportionately. The size of the caudalmost extent of the residual lumen also varies (fig. 8.10E). The pars distalis usually terminates abruptly at its caudal end but sometimes is continuous with the stem remnant.

The pare intermedia develops from the caudal wall of the pouch where the pouch makes contact with the pars nervosa. It is a thin layer of cells at Stage XIX which remains thin in the 4 subsequent stages (fig. 8.10A, B). Lateral processes of the pars intermedia are adjacent to the neural lobe and begin to grow medially, caudal to the neural lobe. The pars tuberalis is a pair of small ridges or of poorly

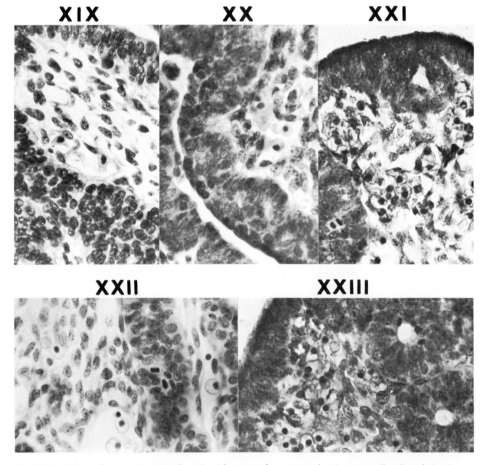

Fig. 8.11. Photomicrographs showing the histogenetic pattern in the pars distalis of the hypophysis in embryos of Stages XIX–XXIII. (×405) Notice the extent of the epithelial ingrowths and the mesodermal vascular elements.

defined small lobes that project rostrally and laterally from the lower, sometimes cupped portion of the adenohypophysis (fig. 8.10*D, E*).

Vascular spaces and angiogenic tissue develop between the cords from meso-dermal elements that invade the region mainly from a rostral direction. The tissue between the cords gradually becomes more vascular and contains a greater variety of cells, some of which become intimately associated with the cord cells (fig. 8.11).

The changes in the pars nervosa are less striking. The lumen that communi-cates with the 3d brain ventricle is gradually lost, and the pars nervosa begins its characteristic folding (fig. 8.10*A, B*). Generally, the level of development of the hypophysis in these baboon stages is slightly more advanced than that reported for man by Streeter (1951).

2. *Pharyngeal hypophysis* (fig. 8.12). A pharyngeal hypophysis (pars pharyn-gea) is present in some of the specimens of Stages XXII and XXIII only. It was observed in almost all the human specimens that Boyd (1956) examined and he suggested it might be more than a mere remnant of the hypophyseal pouch. When present in baboon specimens, it usually has no connection with the stem remnant.

XXII   XXIII

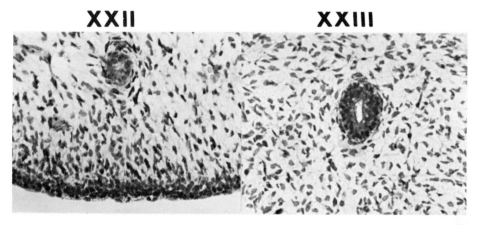

Fig. 8.12. Photomicrographs of transverse sections through the largest part of the most promi-nent pharyngeal hypophysis in embryos of Stages XXII and XXIII. (×230) The pharyngeal hypophysis was found only in these two stages.

At Stage XXII the pharyngeal hypophysis is a narrow cord of cells extending from the pharyngeal epithelium toward the sphenoid cartilage. On transverse sec-tion the nuclei are arranged in a circle and are located near the periphery of the cord with a suggestion of a lumen in the center of the cord. At Stage XXIII the pharyngeal hypophysis varies in size and extent. Sometimes it is a highly organized diverticulum from the pharyngeal roof and extends an impressive distance toward the sphenoid cartilage.

3. *Stem remnant* (fig. 8.13). The manner of absorption of the hypophyseal stem does not differ greatly with that reported for man by Streeter (1951) except that it occurs slightly earlier and its pattern of involution is not as consistent (fig. 8.13). The remnant shows varying degrees of involution which is not always a simple regression. The absorption usually begins just caudal to the sphenoid car-tilage instead of in the middle of the cartilage and absorption may be complete to the pharyngeal attachment. The attachment to the pars distalis then dwindles some-times leaving an isolated segment with variable prominence in the middle of the sphenoid cartilage. The course of the remnant through the cartilage becomes in-creasingly difficult to follow. Most often it appears as a small collection of cells with fusiform nuclei. In all but one of the Stage XXIII specimens, the remnant in

the cartilage is either very faintly visible or has disappeared completely. It is very prominent in one specimen, however, and is isolated in the cartilage with no cranial or caudal continuities.

D. *Vomeronasal Organ* (Fig. 8.14)

A vomeronasal organ in the form of a simple diverticulum as seen in man was never observed, nor was the later stage of a long tapering duct that expands caudally into a blind tubular sac. It is inconsistently and poorly developed in baboon embryos and consequently is not a good criterion by which to classify specimens. At best, it is a thickening in the nasal septal epithelium with a shallow groove or depression on the surface near the center of the placode (see fig. 8.15, *XIX-A*). The greatest amount of indentation was observed in one Stage XXI specimen. Nerve fascicles that are closely associated with fascicles from the olfactory nerve usually

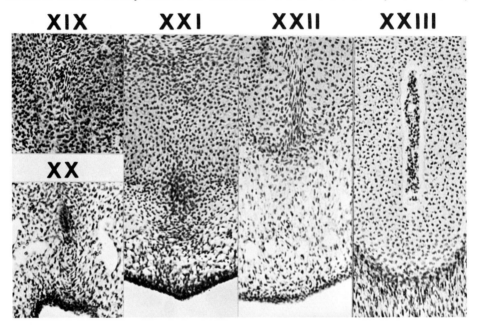

Fig. 8.13. Photomicrographs of the hypophyseal stem remnant at its most prominent level in embryos of Stages XIX–XXIII. (×85) The sections are either at the level of the sphenoid cartilage or just caudal to it.

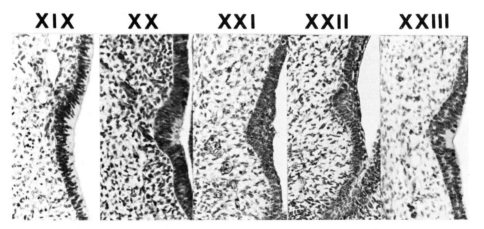

Fig. 8.14. Photomicrographs through the largest part of the most prominent right vomeronasal organ in embryos of Stages XIX–XXIII. (×160)

can be traced to the deep aspect of the placode. These fascicles are less apparent in the Stage XXIII specimens.

E. *Palate and Adjacent Areas* (Figs. 8.15, 8.16)

The lateral palatine processes move from a vertical position lateral to the tongue to a horizontal position above the tongue. By Stage XXIII they begin to fuse with each other and with the nasal septum, separating the nasal from the oral cavity. The nasal conchae become established as prominent projections from the lateral wall of the nasal cavity. Osteoblasts begin to form the premaxilla, maxilla, and palatine bones. Chondroblasts establish the nasal septum and capsule.

The bucconasal membrane has ruptured by Stage XIX allowing the nasal sac to communicate with the oral cavity through the primitive choana (fig. 8.15, *XIX-A*). A lateral extension of primitive palate mesenchyme from beneath the nasal septum to maxillary mesenchyme is not evident. The union between primitive palate and maxillary mesenchyme is accomplished by a medial projection from the rostral portion of the lateral palatine process that will form the secondary palate. Near the tip of the tongue the lateral palatine process is blunted and projects medially: it is closely associated with the midportion of the lateral surface of the tongue (fig. 8.15, *XIX-B*). The process becomes elongated dorsally, points almost caudally, and ends near the linguo-gingival sulcus (fig. 8.15, *XIX-C*). The anlage of the nasal capsule appears as a mesenchymal condensation lateral to the developing conchae (fig. 8.15, *XIX-A, B*). Maxillary osteoblasts can be identified above the labio-gingival lamina and rostral to the primitive choana.

During Stage XX little change occurs in the primitive palate or rostral part of the lateral palatine process. More dorsally the process lengthens and then extends to the junction of the tongue and oral cavity floor (fig. 8.15, *XX-A*). At its dorsal end the process is slender and vertically oriented (fig. 8.15, *XX-B*). In many regions the process and the lateral surface of the tongue come into contact. The area of maxillary osteoblasts increases in size and bone matrix is evident. Differentiation of chondroblasts is apparent in the nasal septum area.

There is essentially no change in the orientation of the process during Stage XXI. Cartilage formation is readily apparent in the nasal septum and cranial portion of the nasal capsule (fig. 8.15, *XXI*). Premaxillary and maxillary bone formation can be distinguished. In the premaxilla, bone forms as a bilateral, horizontal plate that approaches the midline above the incisive portion of the labio-gingival lamina and below the nasal septal cartilage. Maxillary bone forms above the labio-gingival lamina and extends dorsally from a point just rostral to the inferior nasal concha to the superior nasal concha.

The rostral portion of the lateral palatine process projects caudomedially at Stage XXII where the tongue joins the floor of the oral cavity (fig. 8.16, *XXII-A, B*). The process is in close association with the lateral aspect of the tongue. The middle portion of the process has undergone considerable medial rotation and gives the appearance of lifting the tongue (fig. 8.16, *XXII-C*). The dorsal end of the process is directed more caudomedially (fig. 8.16, *XXII-D*). An increase in the distance between the maxillary and mandibular dental laminae suggests that the lower jaw is depressed. The paraseptal cartilage (of the vomeronasal organ) and the frontal process of the maxillary bone are evident (fig. 8.16, *XXII-A*). The middle and inferior conchae are large projections from the lateral wall of the nasal cavity.

By Stage XXIII the rostral two-thirds of the lateral palatine processes have moved to a horizontal position above the tongue, and the epithelia that separate

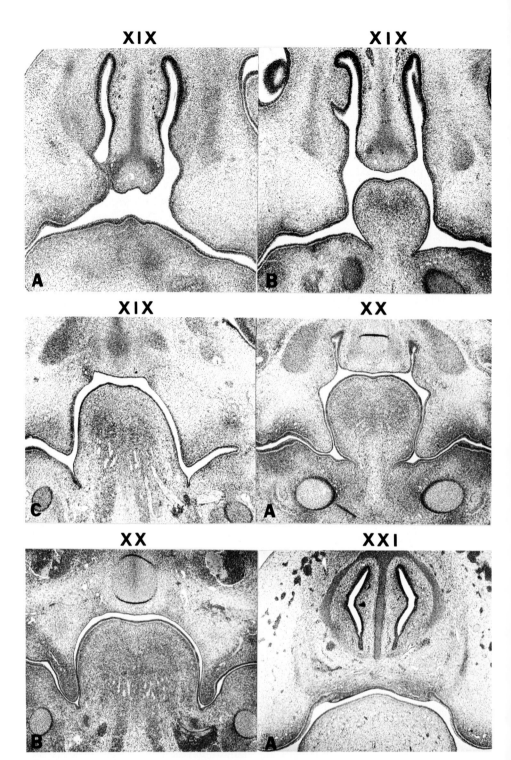

Fig. 8.15. Photomicrographs of coronal sections through various levels of the palate region in representative embryos from Stages XIX, XX, and XXI. (×30) *XIX-A*, section through rostral end of the lateral palatine processes in a Stage XIX embryo; *XIX-B*, section from the same embryo through the processes near the tip of the tongue; *XIX-C*, section from the same embryo through the dorsal portion of the processes; *XX-A*, section through the rostral portion of the lateral palatine processes in a Stage XX embryo; *XX-B*, section from the same embryo through the dorsal portion of the processes; *XXI-A*, section through the primary palate and nasal septum and cartilage in a Stage XXI embryo.

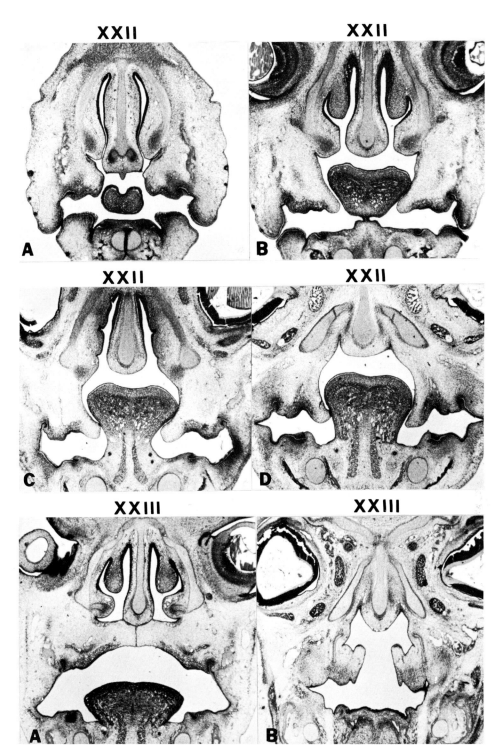

Fig. 8.16. Photomicrographs of coronal sections through various levels of the palate region in representative embryos from Stage XXII and XXIII. (×15) *XXII-A,* section through rostral end of the lateral palatine processes at the tip of the tongue in a Stage XXII embryo; *XXII-B,* section from the same embryo through the rostral portion of the processes where the tongue first joins the floor of the oral cavity; *XXII-C,* section from the same embryo through the middle portion of the processes; *XXII-D,* section from the same embryo through the dorsal portion of the processes; *XXIII-A,* section through the rostral portion of the fused lateral palatine processes in a Stage XXIII embryo; *XXIII-B,* section from the same embryo through the dorsal portion of unfused processes.

the processes from each other and from the nasal septum have fused and are degenerating (fig. 8.16, *XXIII-A*). The lower jaw is depressed further and the tongue appears to be contracted. Both of these actions may be very significant in the removal of the tongue from its earlier position between the processes. At the ossification center of the palatine bone, the processes are in a state of transformation from a vertical to a horizontal position. Soft palate formation in the midline is not complete.

F. *Submandibular Gland* (Fig. 8.17)

Submandibular gland development closely parallels that described in man by Streeter (1951). Angiogenesis appears a little earlier and the manner of lumen formation is slightly different. In man, a suggestion of a lumen in the glandular portion of the duct was observed first followed by a definite lumen in the oral part. In baboon specimens the lumen forms in the middle portion of the main duct first and subsequently spreads to the oral and glandular portions.

At Stage XIX a short, solid, clublike epithelial bud extends from the groove between the tongue and lower jaw (linguo-gingival sulcus) into the margin of an ellipsoidal mass of condensed mesoderm. The mesodermal mass represents the mesenchymal primordium of the gland. The solid epithelial bud is longer at Stage XX and extends to the center of the mesodermal mass, which is denser. Future primary branches appear as knoblike swellings on the bud. A lumen is suggested in the middle segment of the main duct. Angioblastic tissue begins to form. The epithelial bud remains solid and attains a greater length by Stage XXI. It approaches the distal margin of the mesodermal condensation and shows some enlargements with blunt primary branches. An occasional lumen is present in the middle segment of the duct with a suggested lumen at the oral end.

Stage XXII specimens show an epithelial bud with knoblike secondary branches that extend to the periphery of a less compact mesenchymal primordium. Occasional lumina are evident in the primary bud branches, and the oral half of the duct has a definite lumen. The medial aspect of the mesenchymal primordium contains many nerve fibers and is closely related to cells of the submandibular ganglion.

Tertiary and quaternary branches of the bud extend throughout an enlarged mesenchymal primordium at Stage XXIII. A prominent lumen is present throughout the very long main duct. Lumina are distinct in many of the proximal branches of the duct with at least a suggestion of a lumen in the terminal ducts. The mesenchymal primordium is looser, contains much angioblastic tissue, and begins to form a capsule around the periphery of the gland. Ganglion cells remain closely associated to the medial aspect of the mesenchymal primordium, and nerve fibers can be seen coursing between the branching ducts of the gland.

G. *Metanephros* (Figs. 8.18, 8.19)

Previous to Stage XIX the cranial end of the ureteric bud dilates, forming the primitive pelvis of the kidney, and then undergoes the first and second divisions forming the major (primary collecting tubules) and minor (secondary collecting tubules) calyces. The secretory portion of the kidney, which develops from the metanephrogenic mass located at the growing edge of the tubules, begins to differentiate.

By Stage XIX the primordium of the secretory part is condensed as a metanephric cap of tissue surrounding the ampullar ends of the collecting tubules (fig. 8.18). Portions of the metanephrogenic mass in some areas remain undifferentiated. With differentiation, the metanephric caps become first an unorganized cluster of

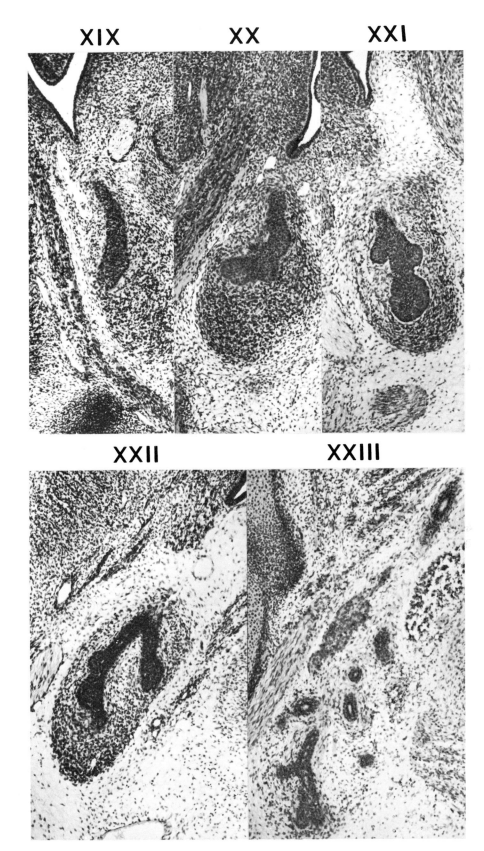

Fig. 8.17. Photomicrographs of sections through the right submandibular gland in embryos of Stages XIX–XXIII. (×80) The tongue and oral cavity are at the top in each photomicrograph.

cells in the angle between the stem and the end of collecting tubules. A lumen then appears within the cell cluster forming a metanephric vesicle.

During Stage XX the collecting tubules increase in number and some of the metanephric vesicles elongate, become S-shaped, make contact with the collecting tubules, and begin to fuse with the collecting tubules (fig. 8.18). Bowman's capsule and the glomerulus can be identified in the lower limb of the S.

The double-walled, spoon-shaped capsule of Bowman is partially invaginated by the glomerulus, and early renal corpuscles form with it at Stage XXI (fig. 8.18). The cleftlike cavity within Bowman's capsule is lined with an outer, parietal layer

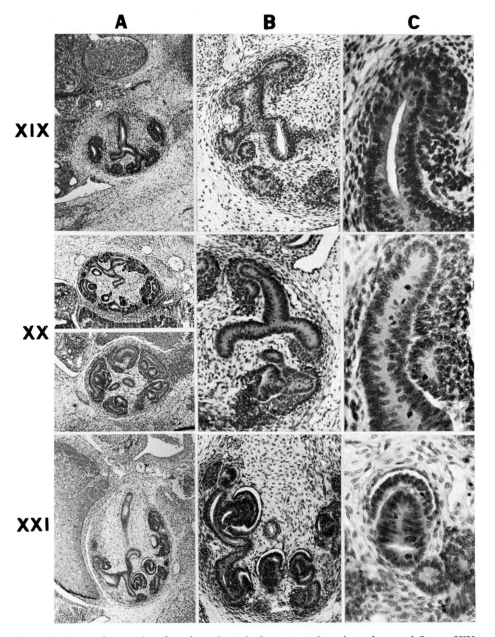

Fig. 8.18. Photomicrographs of sections through the metanephros in embryos of Stages XIX, XX, and XXI showing the origin and development of the kidney tubules. *A*, sections through the middle of the metanephros. (×35) *B, C*, sections through the ends of the collecting tubules and the early renal corpuscle. (*B*, ×90; *C*, ×225)

that is somewhat thinner than the inner, visceral layer adjacent to the glomerulus. The range of development in the kidney becomes greater in successive stages since new tubules are continually being added while those previously established are further differentiated. The most recently developed tubules are peripheral whereas the more advanced ones are nearer the renal pelvis.

Renal corpuscles with larger glomeruli and further differentiated capsules are present in Stage XXII (fig. 8.19). The parietal layer of the capsule consists of

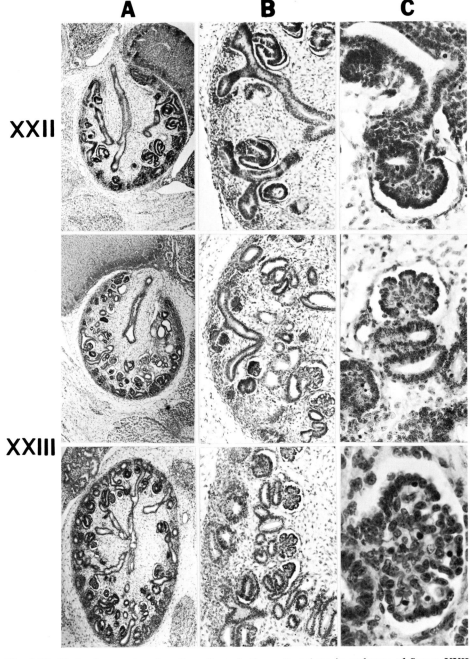

Fig. 8.19. Photomicrographs of sections through the metanephros in embryos of Stages XXII and XXIII showing the development of the kidney tubules. *A*, sections through the middle of the metanephros. (×35) *B, C*, sections through the collecting tubules and renal corpuscle. (*B*, ×85; *C*, ×205, except lower right ×430)

simple squamous epithelium, and the visceral layer is composed of low columnar or cuboidal cells. Differential growth begins to take place in the secretory portion of the tubules.

More collecting and secretory tubules are formed and there is an increased number of large, definitive, renal corpuscles at Stage XXIII (fig. 8.19). The visceral epithelium of Bowman's capsule is predominantly cuboidal but is reduced to a simple squamous layer in the more advanced corpuscles.

H. *Cartilage* (Figs. 8.20, 8.21)

The usefulness of a study of cartilage formation in staging embryos was pointed out by Streeter (1949). He found that cartilage cells in a typical long bone of the extremities (i.e., humerus) pass through an orderly series of transformations. The oldest cartilage cells are located within the growth center in the shaft of the future bone. Adjoining the growth center are successively younger zones or phases of cartilage transformation, the youngest phase (phase 1) being located at the ends of the future bone. Streeter subdivided this orderly process of cartilage transformation into a progression of 5 phases, 1 or more of which may exist simultaneously. The location and extent of the phases represented are constant enough to be used as a criterion for staging.

The following is a general description of the 5 phases of cartilage transfor-

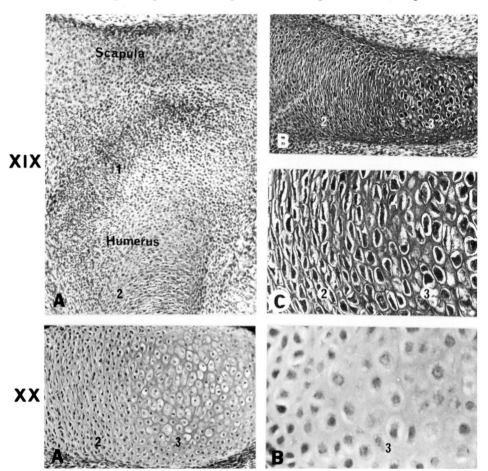

Fig. 8.20. Photomicrographs of sections through the humerus showing the phases of cartilage transformation in embryos of Stages XIX and XX. Sections through phases 1 to 3 of cartilage formation. (*XIX-A, B,* ×110, *C,* ×275; *XX-A,* ×110, *B,* ×275)

mation in the baboon embryonic humerus. Phase 1 cartilage cells are actively proliferating in the surrounding mesoderm with slight enlargement of the older cells (fig. 8.20, *XIX*). The intercellular material is accumulating and the distance between the cells is increasing. Phase 2 cells are elongated and become oriented in rows perpendicular to the long axis of the humerus (fig. 8.20, *XIX, XX*). The cell cytoplasm begins to vacuolize and some of the cells are not yet completely isolated by intercellular material. Phase 3 cartilage cells are more cuboid in shape

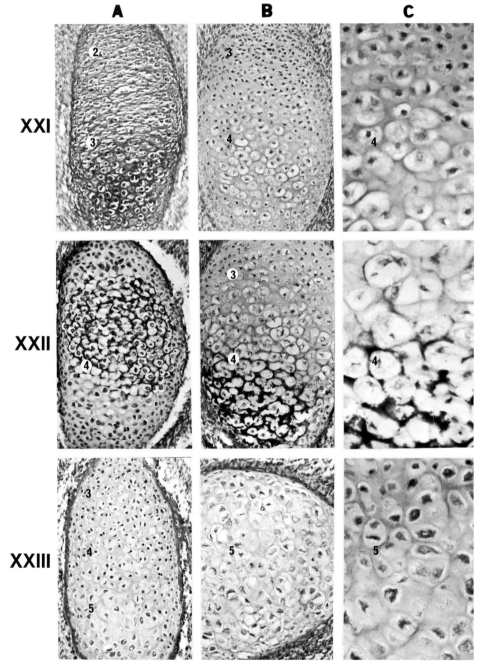

Fig. 8.21. Photomicrographs of sections through the humerus showing the phases of cartilage transformation in embryos of Stages XXI, XXII, and XXIII. Sections through phases 2 to 5 of cartilage formation. (*A, B,* ×100; *C,* ×255)

and enlarge at least threefold (fig. 8.20, *XX*). There is considerable vacuolization of the cytoplasm. During phase 4 the cells attain maximal size and show severe vacuolization (fig. 8.21, *XXI, XXII*). Vacuoles almost completely fill the cytoplasm except for a dense area near the nucleus. The darkly stained intercellular material around the cells gives the cartilage sections a cribriform appearance. Phase 5 cells liquify or disintegrate (fig. 8.21, *XXIII*). The disintegrating cells lose their staining properties and disappear, leaving empty compartments within the matrix.

The humerus in Stage XIX shows the first 3 phases of cartilage cell differentiation (fig. 8.20). Phase 1 cells are at the periphery; deep to them are phase 2 cells which are continuous with the primary growth center that is centrally located and is composed of phase 3 cells. The Stage XX humerus essentially has only an enlarged primary growth center containing more abundant phase 3 cells (fig. 8.20). Phase 4 cells make their initial appearance in the primary growth center at Stage XXI and begin to show disintegration in Stage XXII (fig. 8.21). The darkly stained intercellular matrix characteristic of phase 4 is well demonstrated in Stage XXII. The most notable development during Stage XXIII is cell disintegration which characterizes phase 5 (fig. 8.21).

References

Boyd, J. D. 1956. Observations on the human pharyngeal hypophysis. *J. Endocrin.* 14:66–77.

Streeter, G. L. 1949. Developmental horizons in human embryos: A review of the histogenesis of cartilage and bone. *Contrib. Embryol., Carneg. Inst.* 33:149–67.

———— 1951. Developmental horizons in human embryos: Description of age groups XIX, XX, XXI, XXII, and XXIII. *Contrib. Embryol., Carneg. Inst.* 34:165–96.

9

Placenta

Marshall L. Houston

I. Introduction

The baboon embryo implants superficially and centrally, and chorionic villi form only in the area of original attachment of the uterine mucosa. The result is the formation of a single discoid (sometimes slightly ovoid) placenta with a membranous, avillous chorion peripheral to the placenta and no decidua capsularis. Although certain growth patterns in the early stages result in the formation of a raised margin of fetal and maternal tissue around the periphery of the placenta (fig. 9.8C), which is sometimes referred to as a decidua capsularis incompleta (Noback 1946), this area is soon overgrown and becomes unrecognizable by the 8th week of gestation. The baboon therefore appears to form an intermediate type which implants superficially like the macaque but, like the human, does not form an accessory placenta.

For the presentation of this chapter, the development of the baboon placenta will be divided into 6 periods which will be correlated, when possible, with the stages of embryonic development presented in this volume for comparative purposes. These periods of development are:

A. *Period of the solid trophoblast,* 9th through 11th day of development, during which the blastocyst adheres to the surface of the uterine epithelium, and the trophoblast becomes differentiated into cyto- and syncytiotrophoblast.

B. *Period of the trophoblastic lacunae,* 11th through 13th day of development, characterized by the appearance of irregular clefts in the trophoblastic plate which soon communicate with maternal blood vessels.

C. *Period of villus formation,* 13th through 18th day of development, in which the cytotrophoblast proliferates into solid columns which, in turn, differentiate a mesenchymal core forming chorionic villi.

D. *Period of villus branching and placental angiogenesis,* 18th through 25th day of development, during which the villi undergo primary and secondary branching, and vasoformative elements form within the placental mesenchyme and differentiate into open blood vessels.

E. *Period of the definitive embryonic placenta,* in which structures already present in the placenta undergo further differentiation into the basic forms which they will maintain throughout the remainder of gestation. This process is essentially complete by the 35th day of development.

F. *Period of fetal placental development,* which includes changes occurring during the 2d and 3d trimesters of gestation, such as the disappearance of Langhan's layer, formation of cytotrophoblastic islands and fibrin, and fibrinoid infiltration of the placental tissue.

These periods are, of course, arbitrarily taken on the basis of the more obvious changes occurring at the time; however, the development of many other important structures overlaps these particular periods and there is an inevitable overlap of the day stages due to individual variation between embryos of known ages.

II. Periods of Placental Development

A. *Period of the Solid Trophoblast* (Days 9–11; Embryo Stages IV–V)

During the 9th day of development the baboon blastocyst adheres to the surface of the uterine epithelium (figs. 9.1, 4.1). The trophoblast soon becomes differentiated into cellular and syncytial layers over the area of attachment although no such differentiation is noted in the late preimplantation embryos (see chap. 3). The entire blastocyst is somewhat collapsed, allowing it to flatten out against the

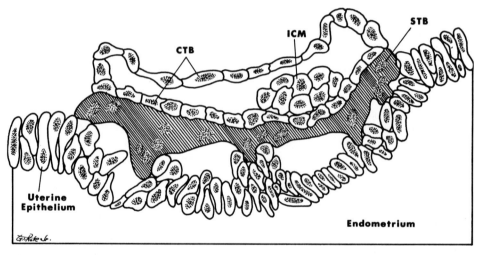

Fig. 9.1. Drawing of a baboon embryo shortly after attachment to the uterine epithelium. The trophoblast in the area of attachment is differentiated into cytotrophoblast (*CTB*) and syncytiotrophoblast (*STB*). Note that the uterine epithelium is not degenerating but is merely disrupted at the points of contact with the trophoblast. *ICM,* inner cell mass. (×550)

uterine epithelium. Actual fusion is not uniform over the entire area of contact but is localized into isolated areas (fig. 9.2). The limiting boundaries of the trophoblastic and uterine epithelial cells remain intact. These isolated areas of fusion are peripheral to the inner cell mass. Alterations of the uterine epithelium are present only at the points of fusion with the trophoblast. Here the nuclei have become disarranged and no longer resemble the tall columnar epithelium of the unfused areas. The cytoplasm in the fused areas is reduced to one-half its normal height, but the nuclei show no pyknosis or other usual signs of degeneration. The uterine stroma underlying the implantation site is edematous and capillaries near the surface are already expanded beyond their normal diameter.

By 10 through 11 days the trophoblast at the site of attachment to the uterine wall has expanded into a heavy plate (fig. 9.3). The greatest portion of this expansion involves the syncytiotrophoblast with the cytotrophoblast reaching no more than two cell layers. Toward the end of this period, small clefts, the precursors of the trophoblastic lacunae, appear within the syncytiotrophoblast of the trophoblastic plate but the clefts remain very narrow and widely separated.

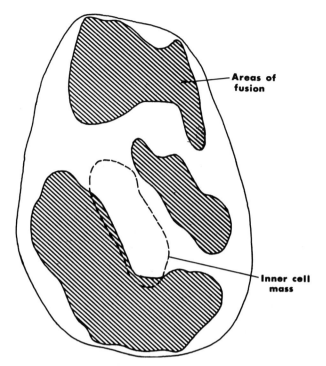

Fig. 9.2. Drawing made from a graphic reconstruction of the implantation site of a 9-day baboon embryo. Fusion is limited to isolated areas within the total area of contact and these areas are located peripheral to the inner cell mass. (×550)

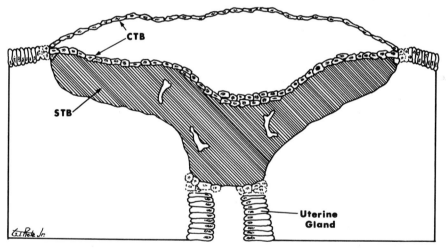

Fig. 9.3. Drawing of a baboon placenta taken in the 11th day of gestation. The syncytiotrophoblast has expanded into a heavy plate and reaches its deepest penetration into the neck of a uterine gland. Small clefts within the syncytium indicate the first signs of the trophoblastic lacunae. The uterine epithelium is completely missing over the area of contact between embryo and endometrium. (×225)

The uterine epithelium is completely missing in the area of implantation but remains intact and unchanged over the remainder of the endometrial surface. The uterine stroma is edematous, with further expanded capillaries. The trophoblast has penetrated only a very short distance into this stroma but expansion is much deeper into the necks of any uterine glands covered by the implantation site.

B. *Period of the Trophoblastic Lacunae* (Days 11–13; Embryo Stage VI)

Open areas within the syncytiotrophoblastic plate become more numerous and rapidly enlarge until by 13 days they occupy a volume almost equal to the trophoblastic tissue itself. Some of these trophoblastic lacunae have anastomosed cen-

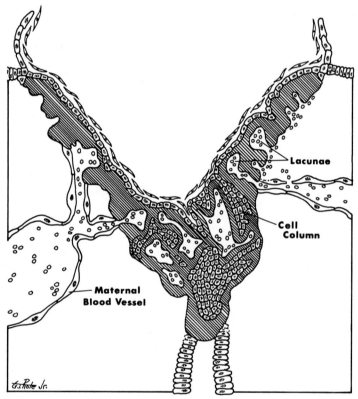

Fig. 9.4. Drawing of a placenta at 13 days gestation showing the large lacunae connecting with maternal blood vessels and the early proliferations of the cytotrophoblast. ($\times$110)

trally but most of them remain isolated (fig. 9.4). The cytotrophoblast is proliferating actively, and columns of cells extend from the embryonic surface of the placenta into the syncytiotrophoblast between the lacunae. Some of these cell columns reach almost to the base of the placenta, and in some cases the axial region near the embryonic surface is loosened indicating very early signs of villus formation (fig. 4.3*A, B*).

The uterine epithelium remains intact peripheral to the implantation site. The trophoblast has expanded directly into the uterine stroma and in doing so has disrupted many of the superficial enlarged capillaries which become continuous with the trophoblastic lacunae, filling them with maternal blood. Leakage of blood into the surrounding tissue spaces is common so that the subplacental area may appear hemorrhagic. The enlargement of the capillaries and superficial venules is limited to the area near the implantation site. The spiral arteries show no signs of alteration.

C. *Period of Villus Formation* (Days 13–18; Embryo Stages VI–VII)

The gross structure of the placenta is changed during this period. The entire trophoblastic mass protrudes above the level of the uterine epithelium and is anchored by a short pedestal of endometrial tissue (figs. 9.5*A*, 9.6). The embryonic surface of the placenta is distinctly convex (figs. 9.6, 4.3*C, D*). The trophoblastic lacunae have begun to anastomose into a continuous open intervillous space, particularly in the central regions of the placenta. The space within the trophoblastic

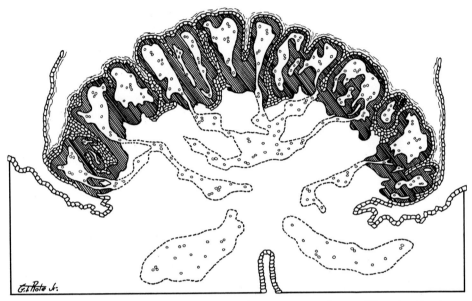

Fig. 9.5. Drawing of a placenta at 16 days showing numerous primary villi and cell columns. Note the raised nature of the placenta and the convex contour of its surface. (×50)

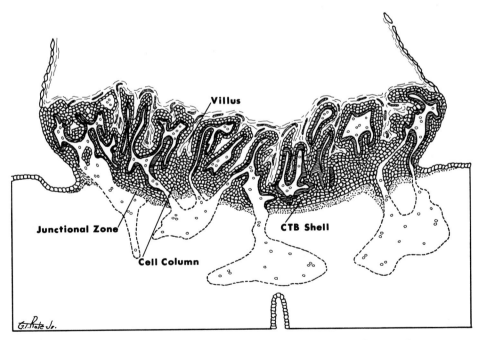

Fig. 9.6. Drawing of a placenta at 19 days gestation. Note the marked increase in cytotrophoblast and the almost complete cytotrophoblastic shell. Chorionic villi show distinct primary branching. (×30)

tissue communicates freely with endometrial sinusoids and thus filled with maternal blood.

The differentiation of the chorionic villi progresses rapidly within the cell columns. Nonbranching, open villi now extend up to half the total depth of the placenta. The villous wall is comprised of an inner layer of evenly arranged, cuboidal cytotrophoblastic cells surrounded externally by a heavy layer of syncytio-trophoblast. The open cores contain scattered mesoblasts which also form a thin layer over the embryonic surface of the placenta.

The cytotrophoblastic cell columns appear to decrease in length because of the more rapid growth of the chorionic villi although there is an increase in their length due to overall placental growth. Cytotrophoblastic cells extend peripherally from the bases of these cell columns and anastomose with similar cells from other cell columns forming an incomplete layer, the cytotrophoblastic shell, at the base

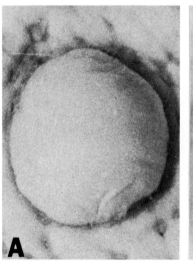
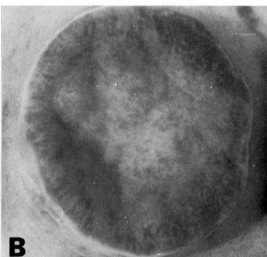

Fig. 9.7. *A*, photograph of a Bouin's fixed implantation site at 16 days gestation with the abpla-cental chorion intact. (×20) (From Houston 1969*a*, fig. 4.) *B*, photograph of a placenta at 20 days gestation with the membranous abplacental chorion still intact. Note the slightly de-pressed central area. (×10) (From Houston 1969*a*, fig. 7.)

of the placenta. The rapid increase in the number of cytotrophoblastic cells has made the syncytiotrophoblast less conspicuous. The cytotrophoblastic shell now marks the point of furthest advancement of the trophoblast against the maternal tissue and the syncytiotrophoblast is now limited to a lining of the intervillous space.

The edema within the uterine stroma is now more generalized. Uterine glands remain conspicuous with no signs of degeneration except where they are in direct apposition with the trophoblast. Expanded vascular channels within the subplacental endometrium communicate freely with the intervillous space, and cytotrophoblastic cells reach their deepest penetration along the walls of these vessels (see fig. 4.3*C, D*). Large granular cells may be found within lumina of the spiral arteries beneath the placenta and some of the fibroblasts immediately surrounding these arteries show early signs of decidualization.

D. *Period of Villus Branching and Placental Angiogenesis* (Days 18–25; Embryo Stages VIII–IX)

The placenta has grown laterally over the surface of the uterine epithelium and is slightly depressed centrally (figs. 9.6, 9.7*B*). This central depression becomes more pronounced during this period until by 25 days the placenta is deeply umbilicated (fig. 9.8*A, B*). The chorionic villi extend almost the entire depth of the placenta, leaving only short cell columns extending to the cytotrophoblastic shell. Most of the villi are distinctly branched. Many primary and secondary branches are evident and near the end of the period tertiary branches are identifiable. Lines of vasoformative mesoblasts are present in the body stalk and on the placental surface near the embryo at 16 days, and by 19 days they have spread throughout the placental mesenchyme (figs. 9.7, 4.6*C*). Open vessels are numerous both on the surface and in the villi at 21 days (fig. 9.8*C*), and nucleated embryonic blood cells may be found in the lumina of these vessels by 25 days.

The cytotrophoblastic shell increases in thickness during this period and forms a uniform, cellular sheath of cytotrophoblast between maternal tissue and the intervillous space (figs. 9.7, 9.8*D*). At the outer margin of the shell, the cytotrophoblastic cells become intermingled with decidual cells, some of which are becoming distinctly necrotic, forming another distinct layer commonly known as the junctional zone. Intercellular space is much less abundant in this layer than in the cytotrophoblastic shell, and because of its degenerating nature, the junctional zone takes on a somewhat heavier stain.

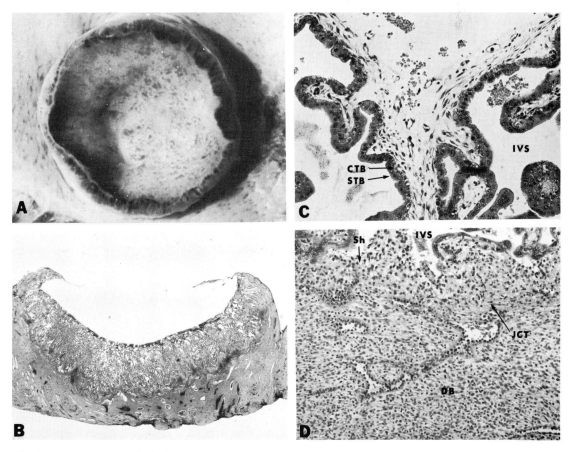

Fig. 9.8. *A*, photograph of a baboon placenta at 24 days. Note the raised margins and the deep umbilication. (×3.5) *B*, photomicrograph of a complete section through a placenta at 25 days showing the complexity of villus branching and the umbilicated nature of the entire placenta. (×4.5) *C*, section through the neck of a chorionic villus at the placental surface at 22 days. Note the double-layered wall of the villus and the open capillaries within the mesenchyme. *CTB*, cytotrophoblast; *IVS*, intervillous space; *STB*, syncytiotrophoblast. (×100) *D*, section through the basal plate of the placenta at 23 days. *Sh*, cytotrophoblastic shell; *JCT*, junctional zone; *DB*, decidua basalis; *IVS*, intervillous space. (×85)

The large venous channels and the spiral arteries of the endometrium communicate with the intervillous space through clefts in the cytotrophoblastic shell (fig. 9.9*A*). The original arteries and veins do not extend beyond the junctional zone. Trophoblastic tissue may be observed along the inner walls of the large vessels beneath the base of the placenta. Scattered large granular cells are found along the course of the spiral arteries and heavy masses of these cells fill the arterial lumen at various points (fig. 9.9*B*). At these points the endothelium of the vessel is attenuated or not recognizable. The muscular walls of the arteries appear normal.

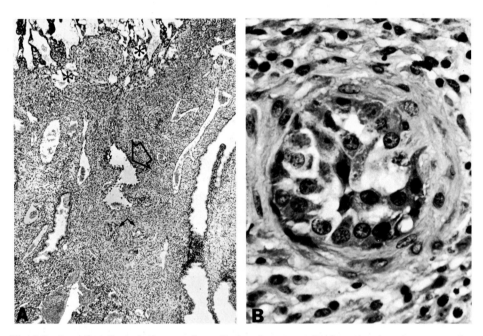

Fig. 9.9. *A*, section through the area beneath the placenta at 23 days showing the course of a spiral artery. Note the abundance of intraluminal cells in some areas (*small arrows*) and the expansion of the vessel into wide clefts nearer the base of the placenta (*large arrow*). Within the cytotrophoblastic shell the channel divides freely so that the contents of a single spiral artery may enter the intervillous space at several points (*). (×30) *B*, transverse section of a spiral artery virtually filled with large intraluminal cells. Note the lack of penetration of these cells into the muscularis. (×460)

The endometrial stroma undergoes rapid decidualization during this period (fig. 9.8*D*). The narrow, spindle-shaped stromal cells become swollen into large, oblong, densely granular cells. The uterine glands appear quite normal beneath the placenta at 18 days even though their upper regions have been destroyed by the invading trophoblast (fig. 4.6). However, by 25 days, signs of degeneration have occurred in the form of a thinning epithelium and a collapse of the glandular walls. However, as with decidualization, these changes are not as marked peripheral to the placenta.

E. *Period of the Definitive Embryonic Placenta* (Days 25–40; Embryo Stages X–XVIII)

The placenta increases steadily in diameter with the margins raised well above the surface of the uterus (fig. 9.10*A*, *B*, *C*). Trophoblastic invasion extends through almost the entire depth of the endometrium. In addition, trophoblast extends peripherally beneath the uterine surface beyond the original area of pene-

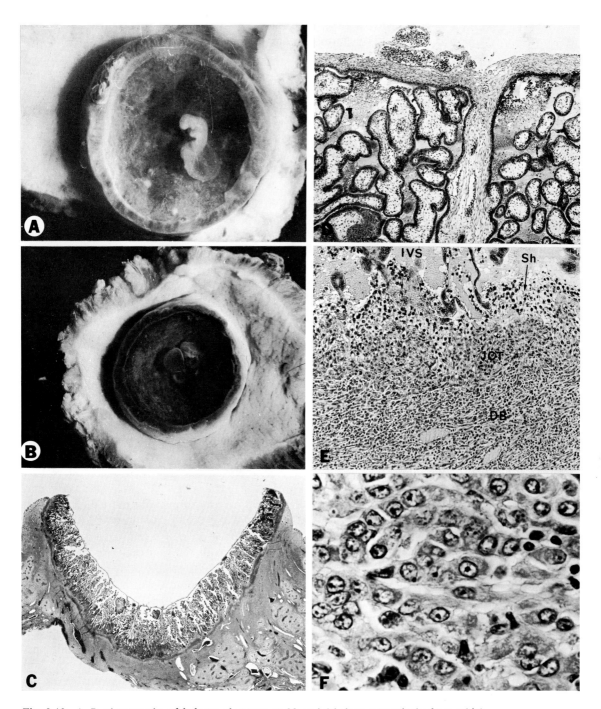

Fig. 9.10. *A, B*, photographs of baboon placentae at 30 and 34 days respectively from which the membranous chorion has been removed. Note the deep umbilication. (*A*, ×3; *B*, ×2) *C*, section through the entire placenta shown in *B*. Note the complex nature of the villus branching and the relatively smooth line of penetration of trophoblastic tissue into maternal tissue. (×4.5) (From Houston 1969a, fig. 17.) *D*, section through the placental surface at 38 days showing the increased density of the mesenchymal stroma and the nature of the embryonic blood vessels on both the surface and in the villi. (×90) (From Houston 1969a, fig. 18.) *E*, section through the basal plate of the placenta at 32 days. Note the increase in intercellular material among the cells of the cytotrophoblastic shell (*Sh*). *DB*, decidua basalis; *IVS*, intervillous space; *JCT*, junctional zone. (×90) *F*, photomicrograph of a typical area of the decidua basalis at 35 days. (×700)

tration. Thus, a rim of tissue is formed in which a layer of decidua overlays an area of attenuated villi, forming what some authors have called the decidua capsularis incompleta (fig. 9.10C).

Branching of the chorionic villi becomes extremely complex during this period and the convolution is greatly increased. While the actual volume of the intervillous space increases steadily with the expansion of the placenta, the ratio of this volume to that of the trophoblastic tissue is decreased by the growth of the trophoblast. In the central regions of the placenta the space is reduced to narrow slits between villous branches (fig. 9.10D). Although the open villi extend to the trophoblastic shell, differentiation of the villous core continues at the villous tips. By 35 days a network of vessels within the chorionic mesenchyme extends over the placental surface and to the extremities of the villi. Vessels within the villi are seldom more than a single endothelial layer, but mesenchymal cells around some of the large vessels on the surface become arranged concentrically around the vessels as an early adventitia (fig. 9.10D).

The cytotrophoblastic shell increases steadily in thickness throughout this period, and elements of intercellular matrix material become more and more evident (fig. 9.10E). The cytotrophoblast and necrosed maternal tissue form a continuous junctional zone as progressive destruction continues deeper into the endometrium. During this period, the first indications appear of a more rapid penetration and destruction of endometrium beneath the main villous masses (fig. 9.10C). This condition leaves ridges of decidua basalis, junctional zone, and cytotrophoblastic shell along the placental base between the main villous trees. These ridges are the beginnings of the intercotyledonary septae. The villous trees form the primordia of the placental cotyledons.

The syncytiotrophoblast changes very little during this period. It lines the intervillous space (fig. 9.10D) and is present to varying degrees as fine strands within the cytotrophoblastic shell. Thus a typical villus consists of an outer covering of syncytiotrophoblast over a single layer of cuboidal cytotrophoblast with an inner core of mesenchyme containing the extraembryonic placental vasculature (fig. 9.10D). This histological relationship persists throughout the villi and on the placental surface.

The remainder of the endometrial surface is unchanged, and although the abembryonic trophoblast is in direct contact with the endometrium opposite the placenta, there is no fusion of the two layers. The uterine glands remain enlarged and degenerate beneath the placenta and have become deflected laterally so that they course diagonally toward the uterine lumen. The entire uterine stroma has become decidualized by 30 days.

Spiral arteries course from the myometrium to the base of the placenta where they open directly into the intervillous space through syncytiotrophoblastic-lined clefts similar to those observed in earlier stages. Venous sinuses drain blood from the intervillous space via direct communications through the junctional zone. No direct connection between arteries and venous sinuses is observed in the decidua basalis at any stage.

F. *Period of Fetal Placental Development* (Days 40–175; Embryo Stages XIX–XXIII)

The placenta is raised well above the surface of the endometrium throughout the fetal period. Specimens from early in the period remain deeply umbilicated but by 100 days have a flat fetal surface, broken only by shallow indentations marking the lobes of the placenta. Both the fetal and maternal surfaces of the placenta show this lobular structure which becomes more pronounced as gestation progresses (fig. 9.11). Each lobe corresponds to a single large villous tree or a combined group of smaller villous trees. Short projections of the basal plate into the intervillous space, the placental septae, seem to be passively formed by a deeper expansion into the endometrium directly beneath the villous tree.

Changes which take place in the placental tissue during the fetal period of development involve changes in structures originally established during the embryonic period of gestation. The amnion only partially fills the chorionic cavity at 50 days, but by 60 days it has expanded and fused with the chorion over its entire inner surface. Although the portion of the amniochorion peripheral to the placenta is in direct contact with the inner surface of the uterus, there is no adhesion or fusion of the two layers until after 100 days of gestation (fig. 9.12*A, B*). At this time the uterine epithelium, which has become progressively thinned, is

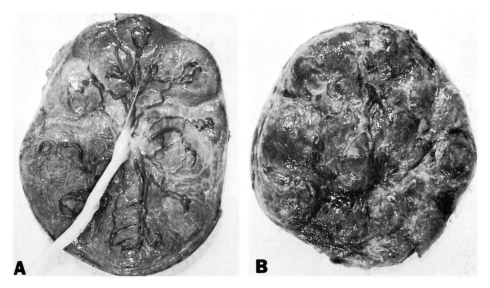

Fig. 9.11. *A*, photographs of the fetal surface and *B*, maternal surface of a full term baboon placenta. Note the central insertion of the umbilical cord and the lobular structure of the placental tissue. (×0.4)

disrupted; and the amniochorion fuses to the underlying stroma to the extent that a heavy layer of decidua parietalis is removed with the amniochorion if they are separated (fig. 9.12*C*). However, no evidence of differentiation into cyto- and syncytiotrophoblast is noted in the abplacental chorion and no villi are formed.

The placenta itself continues to be comprised of the two basic tissues derived early in development, cytotrophoblast and syncytiotrophoblast. During the fetal period, however, the differentiation of these two tissues is different in various areas of the placenta. For convenience in discussing these changes, the placenta may be divided into three basic areas; the *chorionic plate,* the fetal surface of the placenta from which the chorionic villi arise and project distally into the intervillous space; the *chorion frondosum,* or the thick villous area of the placenta; and the *basal plate,* composed of the cytotrophoblastic shell, the junctional zone, and the decidua basalis.

The syncytiotrophoblast continues to line the intervillous space throughout its entire area (figs. 9.13, 9.14, 9.15, 9.16). This layer is characterized by a distinct brush border on its distal surface throughout the fetal period.

Cytotrophoblast steadily proliferates in two of the three areas of the placenta, the chorionic plate and cytotrophoblastic shell. At 45 days the cytotrophoblast of the chorionic plate consists of a single layer of polyhedral, chromophobic cells which stand out in clear contrast to the deeply staining syncytiotrophoblast (fig. 9.13*A*). By 100 days the cytotrophoblast has begun to proliferate in localized areas, forming heavy patches of cells over the placental surface (fig. 9.13*B*). This proliferation soon becomes generalized, and by 120 days this layer is thickened overall, reaching up to 8 cell layers by term (fig. 9.13*C*). The cytotrophoblastic shell at 45 days already consists of 4–8 layers of loosely arranged cells (fig. 9.14). The thickness of this layer gradually increases and varies considerably from one area to another. The cells at the distal margin of the shell become intermingled with necrotic maternal cells and remnants of cells forming the junctional zone of the basal plate.

One of the most notable characteristics of the fetal period is the accumulation of noncellular matrix materials in the two above mentioned areas of the placenta.

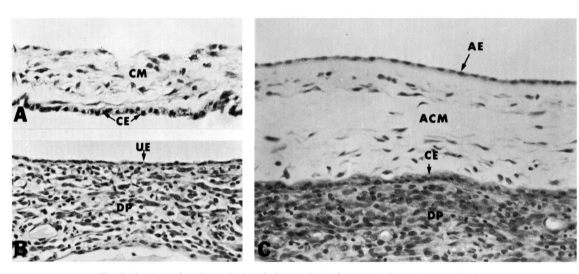

Fig. 9.12. *A*, section through the abplacental chorion at 50 days. *CE*, chorionic ectoderm; *CM*, chorionic mesoderm. (×200) *B*, section through the decidua parietalis (*DP*) at 50 days showing the intact but attenuated uterine epithelium (*UE*). (×200) *C*, section through the fused amnio-chorion and decidua parietalis at 120 days gestation. The uterine epithelium is completely missing and the chorionic ectoderm (*CE*) is fused directly to the underlying decidua parietalis (*DP*). *AE*, amniotic ectoderm; *ACM*, amnio-chorionic mesoderm. (×200)

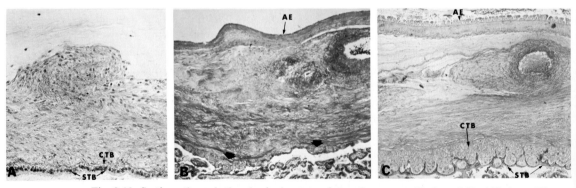

Fig. 9.13. Sections through the chorionic plate of the placenta at 50 days (*A*), 107 days (*B*), and 170 days (*C*). The inner surface of the plate is lined by syncytiotrophoblast (*STB*) throughout. The cytotrophoblast (*CTB*) is a single layer early in the fetal period (*A*) but proliferates in certain areas by 100 days (*B, arrows*) and is generally thickened by term (*C*). *AE*, amniotic ectoderm. (*A*, ×85; *B*, ×35; *C*, ×75)

The mesenchymal stroma over the surface of the chorionic plate becomes increasingly heavy by the addition of collagenous fibers (fig. 9.13). In addition, noncollagenous fibrin and fibrinoid materials accumulate among the cytotrophoblastic cells of the chorionic plate. In the cytotrophoblastic shell, very little collagenous material is present, but fibrin and fibrinoid accumulation is extremely heavy (fig. 9.14). The distribution of this material is homogenous except for a heavy layer along the base of the shell bordering the junctional zone. In this area fibrin is deposited in heavy bands forming a layer known as Nitabuch's membrane, which is first recognized at about 100 days gestation.

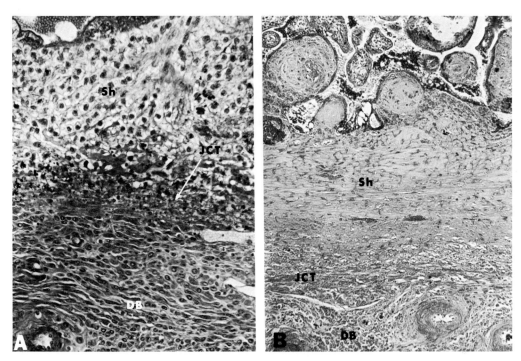

Fig. 9.14. Sections through the basal plate of the placenta at 50 days (*A*) and 167 days (*B*) showing the cytotrophoblastic shell (*Sh*), junctional zone (*JCT*), and decidua basalis (*DB*). Note the tremendous increase in the amount of intercellular matrix within the cytotrophoblastic shell. (*A*, ×160; *B*, ×80)

The chorion frondosum becomes more and more complex as branches of the chorionic villi become more numerous. Free terminal villi arise as fine extensions or "buds" of the villous wall from all levels of the villous tree. Although these terminal villi are continually added, the frequency of new villous buds gradually diminishes toward term.

Within the chorionic villi, changes in the cytotrophoblast are very different than those noted in the chorionic plate and cytotrophoblastic shell. At 45 days, the villous wall remains comprised by an outer layer of syncytiotrophoblast and an inner layer of cytotrophoblast, Langhan's layer (fig. 9.15*A*). However, by 100 days Langhan's layer is discontinuous (fig. 9.15*B*), and by term only scattered cytotrophoblastic cells remain in the villi (fig. 9.15*C*). This leaves a villous wall composed of a single layer of syncytiotrophoblast (fig. 9.16). This syncytial wall is thinner where embryonic capillaries lie against its inner surface, and the nuclei clump together away from these thinner areas.

The decidua basalis is permeated by large venous sinusoids which communicate freely with the intervillous space through syncytiotrophoblastic lined clefts of various sizes. Spiral arteries communicate directly with the intervillous space through similar clefts which in some cases appear to form a labyrinth within the cytotrophoblastic shell, connecting at only one point with the spiral artery but emptying into the intervillous space at several points. Large, intra-arterial cells, prominent in early stages of placentogenesis, are reduced in number by 45 days and are seldom present after 100 days. The cells of the uterine stroma are distinctly decidualized throughout the fetal period, and the decidua basalis becomes increasingly more dense toward term.

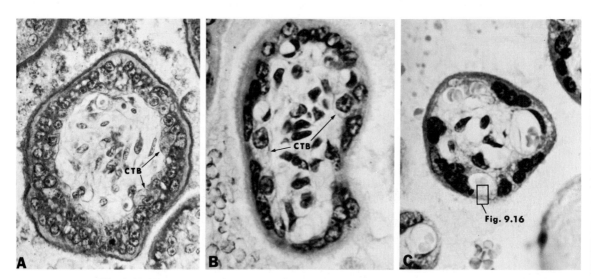

Fig. 9.15. Transverse sections through chorionic villi at 45 days (*A*), 103 days (*B*), and term (*C*). The cytotrophoblastic layer (*CTB*) is complete at 45 days, incomplete at 100 days, and absent at term. Note the migration of the embryonic capillaries to the villous wall and the thinning of the wall with the nuclei grouped away from these points of contact (*C*). (*A*, ×350; *B*, ×425; *C*, ×550) (From Houston 1969*b*, fig. 8.)

III. Placental Circulation

A. *Early Development of Umbilical Vessels*

The umbilical (allantoic) arteries are distinct at each side of the allantois by the 21st day of gestation, and a narrow connecting vessel is identifiable in most specimens. These umbilical vessels arise in the embryo as direct continuations of the paired dorsal aortae but by 30 days are found extending from each common iliac artery just distal to the bifurcation. At this time the connection between the two may be in the form of a short vessel or a direct anastomosis (fig. 9.17). A connection may be identified between 21 and 50 days in approximately 85% of the embryos.

B. *Transverse Communicating Artery*

By term, the body stalk and its contained vessels have developed into an elongated umbilical cord which measures about 25–30 cm in length. A transverse communicating artery is identifiable in 70% of the term placentae, significantly less than the 85% noted in the earlier specimens. Those communicating vessels found in the term placentae may be classified into 5 categories (fig. 9.18).

Type I. A direct connection between the umbilical arteries, in the form of a short vessel nearly equal in size to the umbilical arteries themselves, or a short anastomosis of the two umbilical arteries.

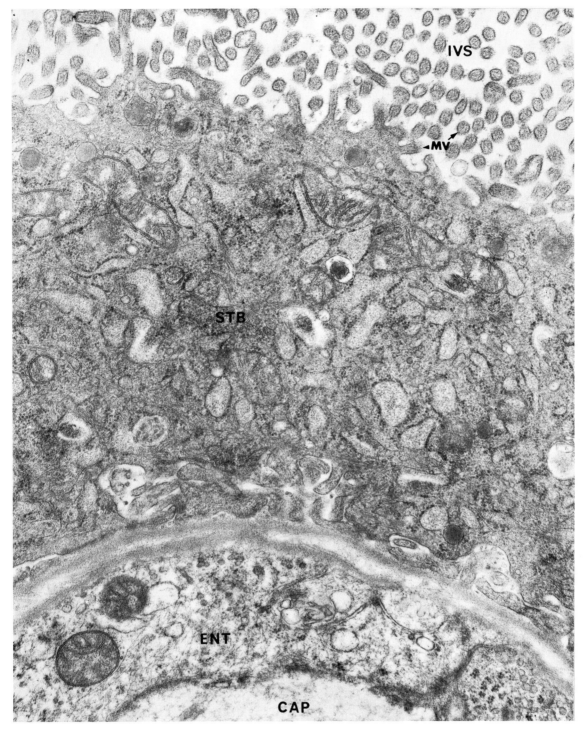

Fig. 9.16. An electron micrograph of a section through the wall of a chorionic villus at term such as indicated in fig. 9.15C. Note that the wall is a single layer of syncytiotrophoblast (*STB*) covered on its outer surface by microvilli (*MV*) and directly adjacent to an embryonic capillary (*CAP*) on its internal border. *IVS*, intervillous space; *ENT*, endothelial cell of the embryonic capillary. (×27,500)

*Type I*a. A communicating vessel as in Type I but with one or more branches of varying size and complexity arising from this vessel.

Type II. A direct communicating vessel between a major branch of one umbilical artery and a similar branch of the other artery. This type of connection is always in the form of a distinct vessel with no direct anastomoses observed between branches of the umbilical arteries. No specimens are observed in which the connection is between branches of the same umbilical artery.

*Type II*a. A communication between major branches of the umbilical arteries as in Type II but with one or more branches of varying size and complexity arising from the transverse communicating vessel itself.

Type III. A communication similar to Types I*a* or II*a* but with a markedly unequal portion from one of the umbilical arteries.

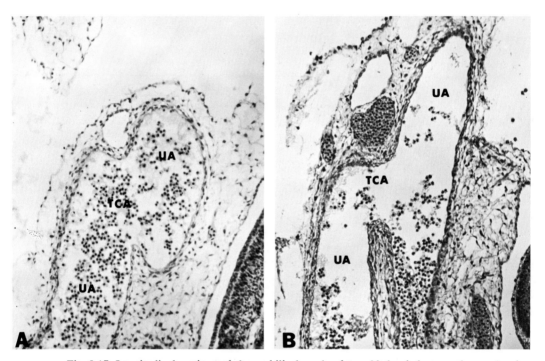

Fig. 9.17. Longitudinal sections of the umbilical cords of two 30-day baboon embryos showing the early transverse communicating artery (*TCA*) between the paired umbilical arteries. (×160) (From Houston and Hendrickx 1968, fig. 1.)

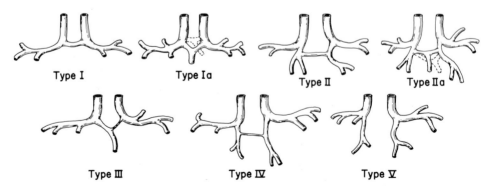

Fig. 9.18. Illustrations of the different types of transverse communicating arteries found in the baboon. (From Houston and Hendrickx 1968, fig. 3.)

Type IV. A communication similar to Types I or II but only a very minute vessel, capable of only limited blood flow.

Type V. No communication.

TABLE 9.1

INCIDENCE OF TYPES OF TRANSVERSE
COMMUNICATING ARTERIES

Type	%	Type	%
I	20.8	III	1.6
Ia	15.2	IV	0.8
II	13.6	V	29.6
IIa	10.4	Other	8.0

C. *Branching Patterns of Placental Vessels*

The umbilical cord enters the placenta within 2 cm of the center in approximately 95% of the cases. Upon reaching the placenta the two umbilical arteries and the single umbilical vein divide into several branches which extend radially over the surface of the placenta. Arterial and venous branches may be distinguished by the fact that the veins are of slightly larger diameter than the arteries and invariably course beneath the arteries where they cross one another.

Two basic branching patterns are found in the placental arteries. The magistral pattern (fig. 9.19*A*) is the most common (66%): the major branches course almost undiminished toward the margin of the placenta, giving off smaller branches into the chorionic plate along the entire length. The remaining 34% display a disperse pattern (fig. 9.19*B*) in which the major arterial branches divide dichotomously several times, diminishing in caliber each time so that only fine vessels reach the placental periphery. The pattern of the placental veins closely corresponds with the pattern of the arteries which they accompany.

In most cases, the branches of each umbilical artery supply approximately an equal area of the placenta (fig. 9.19*A*, *B*). However, in approximately one-fourth of the cases one artery gives off branches to two-thirds or more of the placenta while the other artery supplies only the remainder (fig. 9.19*C*). Also it is not uncommon to find each umbilical artery supplying only about one-third of the placenta (symmetrical) and branches of a large transverse communicating artery supplying the remainder (fig. 9.19*D*).

Branches from the surface vessels of the placenta extend into the necks of the large villous stems and subdivide to vascularize the entire villous tree. The tunica media and adventitia of the surface vessels are lost soon after entering the villous stem and the remainder of the vessel is comprised only of a single endothelial layer. The villous vessels course axially within the main villous stems and their major branches. However, within the smaller terminal villi, the many capillaries become progressively more peripherally located until by term they lie directly against the thinner, inner wall of the syncytiotrophoblast.

IV. Comparison and Classification

The developmental sequence and structure of the baboon placenta are very similar to man and the rhesus monkey, the only other higher primates in which this information is well documented. However several basic differences are worthy of note. Most evident of these is the fact that the baboon forms a single discoid placenta very similar to the human. A double discoid placenta, the "normal" condition in the rhesus monkey, occurs in far less than 1% of the cases, no more often than in man.

The baboon blastocyst implants superficially, and further growth of the embryo and placenta takes place without penetrating completely beneath the surface of the endometrium as in the interstitial implantation of man. This superficial implantation is also common to the rhesus monkey. However, immediately after implantation, the uterine epithelium of the rhesus monkey begins to proliferate actively to the extent that a heavy "epithelial plaque" is formed beneath the developing placenta which persists through the 17th day of development. No such epithelial proliferation occurs in the baboon or in man.

The process of decidualization differs markedly in all three species. In man the cells of the uterine stroma begin to become swollen and transformed into typical decidual cells almost immediately after implantation, whereas in the rhesus

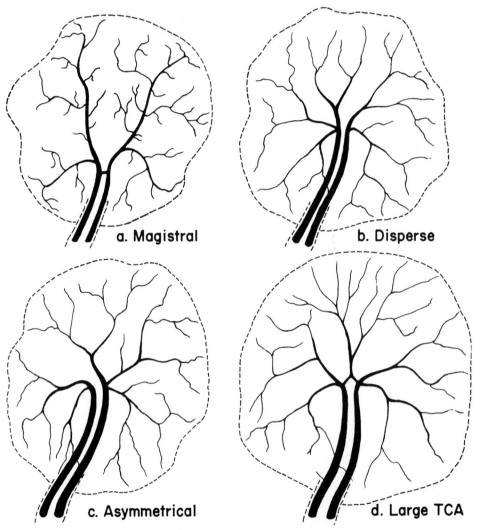

Fig. 9.19. Illustration to show branching patterns and vessel distribution in the baboon placenta: (*a*) magistral, (*b*) disperse, (*c*) asymmetrical, (*d*) fully one-third of the placenta supplied by branches of the transverse communicating artery (*TCA*). Both *a* and *b* also demonstrate a symmetrical distribution of the placental vessels. (From Houston and Hendrickx 1968, fig. 2.)

monkey only a very few stromal cells become hypertrophied later in gestation, a condition which Ramsey and Harris (1966) interpret as a complete lack of a decidual reaction. The baboon, however, seems to form an intermediate condition with a much more prominent decidual reaction than is found in the rhesus monkey but one which forms later in gestation and does not become as dense as in man.

Other characteristics of placental development are amazingly similar between the three species. This is particularly true of the development and final structure of the chorionic villi and the relationship established between embryonic and maternal blood. Thus in the actual functional unit of the placenta, the point of exchange between mother and offspring, very little difference in morphology exists between man, rhesus monkey, and baboon.

By the Grosser method of placental classification, the term baboon placenta reaches a stage of development which places it within the villous hemochorial group. With the aid of electron microscopic observations, it is possible to determine that the outer syncytial wall of the free villi is composed of a single layer of syncytiotrophoblast (fig. 9.16), thus further classifying it as a hemomonochorial placenta using the criteria of Enders (1965).

References

Bacsich, P., and C. F. V. Smout 1938. Some observations on the fetal vessels of the human placenta with an account of the corrosion technique. *J. Anat.* 72: 358–64.

Boyd, J. D., and W. J. Hamilton 1967. Development and structure of the human placenta from the end of the 3rd month of gestation. *J. Obstet. Gynaecol. Brit. Commonw.* 74:161–226.

Coventry, A. F. 1923. The placenta of the Guinea baboon. *Anat. Rec.* 25:237–55.

Enders, A. C. 1965. A comparative study of the fine structure of the trophoblast in several hemochorial placentas. *Amer. J. Anat.* 116:29–68.

——— 1968. Fine structure of anchoring villi of the human placenta. *Amer. J. Anat.* 122:419–52.

Gilbert, C., and C. H. Heuser 1954. Studies in the development of the baboon (*Papio ursinus*): A description of two presomite and two late somite stage embryos. *Contrib. Embryol., Carneg. Inst.* 35:11–54.

Hamilton, W. J., and J. D. Boyd 1960. Development of the human placenta in the first 3 months of gestation. *J. Anat.* 94:297–328.

Hillemann, H. H. 1955. *Organization, histology and circulatory pattern of the near-term placenta of the Guinea baboon,* Papio cynocephalus, *Demarest.* Corvallis: Oregon Agricultural College Press.

Hertig, A. T. 1935. Angiogenesis in the early human chorion and in the primary placenta of the macaque monkey. *Contrib. Embryol., Carneg. Inst.* 25;37–82.

Hertig, A. T.; J. Rock; and E. C. Adams 1956. A description of 34 human ova within the first 17 days of development. *Amer. J. Anat.* 98:435–94.

Heuser, C. H., and G. L. Streeter 1941. Development of the macaque embryo. *Contrib. Embryol., Carneg. Inst.* 29:15–55.

Houston, M. L. 1969a. The villous period of placentogenesis in the baboon (*Papio* sp.). *Amer. J. Anat.* 126:1–16.

——— 1969b. The development of the baboon (*Papio* sp.) placenta during the fetal period of gestation. *Amer. J. Anat.* 126:17–30.

Houston, M. L., and A. G. Hendrickx 1968. Observations on the vasculature of the baboon placenta (*Papio* sp.) with special reference to the transverse communicating artery. *Folia Primatol.* 9:68–77. Basel and New York: S. Karger.

Noback, C. R. 1946. Placentation and angiogenesis in the amnion of a baboon (*Papio papio*). *Anat. Rec.* 94:553–67.

Ramsey, E. M. 1949. The vascular pattern of the endometrium of the pregnant rhesus monkey (*Macaca mulatta*). *Contrib. Embryol., Carneg. Inst.* 33:113–47.

Ramsey, E.M., and J. W. S. Harris 1966. Comparison of uteroplacental vasculature and circulation in the rhesus monkey and man. *Contrib. Embryol., Carneg. Inst.* 38:61–70.

Starck, D. 1956. Primitiventwicklung und Plazentation der Primaten. In H. Hofer, A. H. Schultz and D. Starck, *Primatologia,* 1:723–886. Basel and New York: S. Karger.

Wislocki, G. B., and G. L. Streeter 1938. On the placentation of the macaque (*Macaca mulatta*), from the time of implantation until the formation of the definitive placenta. *Contrib. Embryol., Carneg. Inst.* 27:1–66.

Wislocki, G. B., and H. S. Bennett 1943. The histology and cytology of the human and monkey placenta, with special reference to the trophoblast. *Amer. J. Anat.* 73:335–449.

Appendix: Fetal Growth

Andrew G. Hendrickx/Marshall L. Houston

The gestation period of the baboon averages 26 weeks in length (Kriewaldt and Hendrickx 1968) and can be divided conveniently into two shorter periods, the embryonic period, which consists of the first 7 weeks of development, and the fetal period, which consists of the remaining 19 weeks. The stages of embryonic development are presented in chapters 3–8 while body and organ weights as well as selected body measurements are presented here as an indication of fetal growth. Numerical values and statistical data are presented in tables A.1 and A.2. The same data are presented in graphic form (figs. A.1 to A.15) to aid the investigator in comparing body and organ weights (figs. A.1 to A.10) and body measurements (figs. A.11 to A.15). The data on embryonic crown-rump (CR) length (fig. A.16), chorionic diameter (fig. A.17) and estimated fertilization age (figs. A.18 and A.19) are included to provide a means of comparing these characteristics between and within embryonic stages.

A total of 65 fetuses were weighed and measured, although all weights and measurements were not made on each specimen. The fetuses were nearly equally divided as to sex: 32 males and 29 females (4 were too immature for sex to be determined without microscopic examination of the gonads). The fetuses were recovered by hysterotomy or natural abortion between the 8th and 26th week of gestation. Only the intact nonautolyzed aborted fetuses were used. Their ages were determined as described in chapter 2. All measurements were made according to the technique described by Schultz (1929).

173

TABLE A.1

Body and Organ Weights for the Fetal Period

(In Grams)

	Age in Weeks									
	8–9	10–11	12–13	14–15	16–17	18–19	20–21	22–23	24–25	26
Fetus										
(fresh)	(N=5)	(N=4)	(N=6)	(N=3)	(N=4)	(N=6)	(N=8)	(N=5)	(N=4)	...
Range {	3.75	15	55	111	220	212	370	495	650	...
	13.7	25	150	209	269	364	636	680	800	...
Mean	7.5	21.7	83.7	165	240	308	471	578	737	...
SD	5.06	4.58	34.79	49.67	20.68	46.22	92.54	77.23	62.83	...
SE	2.27	2.29	14.20	28.67	10.33	18.87	32.71	34.53	31.42	...
Brain										
(fixed)	...	(N=3)	(N=6)	(N=2)	(N=2)	(N=6)	(N=8)	(N=7)	(N=3)	...
Range {	...	2.44	2.85	9.25	19.30	24.80	41.00	50.16	70.82	...
	...	9.50	11.66	12.65	33.20	53.60	61.50	84.60	86.76	...
Mean	...	4.90	6.99	10.90	26.30	41.40	51.81	63.42	77.26	...
SD	...	3.99	2.94	2.49	9.83	12.21	8.59	14.03	8.40	...
SE	...	2.30	1.20	1.76	6.95	4.98	3.03	5.29	4.85	...
Thyroid										
(fresh)	...	(N=3)	(N=3)	(N=3)	(N=2)	(N=5)	(N=6)	(N=4)	(N=4)	...
Range {	...	.001	.020	.025	.040	.040	.060	.100	.120	...
	...	.035	.030	.075	.121	.092	.220	.380	.220	...
Mean	...	.019	.023	.056	.080	.070	.132	.185	.178	...
SD	...	.016	.007	.028	.057	.021	.063	.130	.042	...
SE	...	.009	.004	.016	.040	.003	.026	.066	.020	...
Thymus										
(fresh)	...	(N=3)	(N=6)	(N=2)	(N=2)	(N=5)	(N=6)	(N=5)	(N=4)	...
Range {	...	.02	.05	.17	.61	.49	.32	.45	.94	...
	...	.08	.62	.45	2.83	1.17	2.81	1.70	3.35	...
Mean	...	.04	.22	.31	1.72	.70	1.50	1.11	2.44	...
SD	...	.032	.21	.32	1.57	.27	.93	.59	1.05	...
SE	...	.019	.09	.23	1.11	.12	.38	.26	.53	...
Heart										
(fresh)	...	(N=3)	(N=6)	(N=3)	(N=3)	(N=5)	(N=6)	(N=4)	(N=3)	...
Range {	...	.14	.39	.79	1.37	2.18	2.29	3.66	3.70	...
	...	.16	.81	1.60	2.16	2.69	5.79	7.25	5.71	...
Mean	...	.15	.55	1.30	1.65	2.33	3.79	4.96	4.54	...
SD	...	.12	.16	.44	.44	.21	1.46	1.66	1.03	...
SE	...	.007	.07	.26	.23	.09	.60	.83	.60	...
Lung										
(fresh)	...	(N=3)	(N=6)	(N=2)	(N=3)	(N=5)	(N=6)	(N=3)	(N=3)	...
Range {	...	.23	1.80	3.46	6.33	7.31	5.34	7.23	13.85	...
	...	.83	5.83	6.23	7.50	10.85	15.75	21.23	20.87	...
Mean	...	.51	2.88	4.85	6.80	8.84	11.20	14.90	16.67	...
SD	...	.31	1.48	1.96	.59	1.48	4.62	7.07	3.70	...
SE	...	.17	.60	1.38	.34	.67	1.88	4.09	2.14	...
Liver										
(fresh)	...	(N=2)	(N=6)	(N=3)	(N=3)	(N=5)	(N=6)	(N=5)	(N=3)	...
Range {	...	.73	1.91	3.69	7.42	6.07	5.44	5.09	21.14	...
	...	.77	4.77	6.21	10.26	11.70	28.75	36.36	26.26	...
Mean	...	.75	2.79	5.22	8.38	9.83	15.12	16.54	23.67	...
SD	...	...	1.05	1.35	1.63	2.37	8.08	11.82	2.55	...
SE	...	...	.44	.78	.93	1.06	3.29	5.29	1.47	...
Spleen										
(fixed)	...	(N=1)	(N=2)	(N=1)	(N=2)	(N=6)	(N=3)	(N=3)	(N=2)	...
Range {	...	...	.07	...	.41	.28	.43	.80	1.83	...
	...	...	.10	...	1.13	1.12	.75	1.23	2.23	...
Mean	...	...	.085	...	.77	.60	.60	1.28	2.03	...
SD	...	...	.022	...	.51	.28	.16	.50	.28	...
SE	...	...	.016	...	.36	.11	.09	.28	.20	...

					Age in Weeks					
	8–9	10–11	12–13	14–15	16–17	18–19	20–21	22–23	24–25	26
Spleen (fresh)	...	(N=3)	(N=5)	(N=3)	(N=2)	(N=5)	(N=6)	(N=2)	(N=3)	...
Range {	...	.02	.11	.26	.37	.28	.38	.88	1.53	...
	...	.04	.28	.60	1.00	.82	1.83	1.40	2.56	...
Mean	...	.03	.16	.42	.68	.60	1.05	1.14	1.95	...
SD	...	.01	.07	.17	.45	.20	.51	.37	.54	...
SE	...	.004	.03	.10	.32	.09	.21	.27	.10	...
Adrenal (fresh)	...	(N=3)	(N=4)	(N=2)	(N=2)	(N=4)	(N=5)	(N=5)	(N=3)	...
Range {	...	.04	.10	.10	.09	.05	.11	.07	.24	...
	...	.05	.13	.12	.15	.19	.62	.38	.32	...
Mean	...	.047	.107	.11	.12	.14	.24	.26	.28	...
SD	...	.007	.013	.014	.042	.062	.22	.122	.04	...
SE	...	.004	.007	.010	.030	.031	.10	.055	.022	...
Kidney (fresh)	...	(N=2)	(N=6)	(N=2)	(N=2)	(N=5)	(N=4)	(N=3)	(N=3)	...
Range {	...	.08	.40	.79	1.52	1.82	1.48	2.15	3.25	...
	...	.15	1.53	1.24	2.40	2.08	3.25	4.04	4.07	...
Mean	...	.115	.70	1.01	1.96	1.96	2.46	3.02	3.68	...
SD	...	.050	.43	.32	.62	.114	.86	.95	.41	...
SE	...	.035	.17	.02	.44	.051	.14	.55	.24	...

TABLE A.2
Body Measurements for the Fetal Period
(In mm)

	8–9	10–11	12–13	14–15	16–17	18–19	20–21	22–23	24–25	26
Sitting height (fixed)	(N=6)	...	...	(N=1)	(N=3)	(N=2)	(N=3)	(N=4)	(N=2)	(N=2)
Range	26.5–59.7	...	...	...	140–273	180–217	151–226	208–240	218–220	215–215
Mean	44.2	...	...	...	187.6	198.5	192.3	223.5	219	215
SD	12.73	...	...	...	74.07	26.17	38.15	14.61	1.41	...
SE	5.20	...	...	...	42.77	18.50	22.02	7.31	1.00	...
Sitting height (fresh)	(N=5)	(N=3)	(N=6)	(N=3)	(N=3)	(N=7)	(N=10)	(N=8)	(N=3)	...
Range	28.0–55.7	62.9–77.0	89.5–138.0	120–141	146–152	149.9–190.0	151.0–226.4	200–240	210–225	...
Mean	43.6	70.3	107.7	133.7	149.6	171.7	190.7	212.5	216.6	...
SD	10.86	7.07	14.85	11.83	3.16	13.23	22.11	14.21	7.65	...
SE	4.86	4.09	6.64	6.84	1.82	5.00	6.99	5.00	4.42	...
Standing height (fresh)	...	...	(N=1)	...	(N=2)	(N=4)	(N=6)	(N=4)	(N=3)	...
Range	...	...	...	...	237.6–254.5	252.6–305.3	313.9–380.3	242.8–355.8	372.3–384.8	...
Mean	...	...	...	...	246.0	276.4	335.4	350.8	377.4	...
SD	...	...	...	...	11.96	20.88	25.49	11.40	10.89	...
SE	...	...	...	...	8.46	10.44	10.41	5.69	6.28	...
Hand length (fixed)	(N=2)	...	...	(N=1)	(N=1)	(N=2)	(N=4)	(N=4)	(N=2)	(N=2)
Range	3.0–8.6	...	...	...	...	34.2–45.5	31–45	54–62	47.7–51.2	49–56
Mean	5.82	...	...	...	...	39.9	41.2	56.75	49.5	52.5
SD	3.96	...	...	...	...	7.99	6.86	3.61	2.49	4.90
SE	2.80	...	...	...	...	5.65	3.43	1.80	1.76	3.46
Hand length (fresh)	...	...	(N=2)	(N=1)	(N=2)	(N=6)	(N=6)	(N=6)	(N=3)	...
Range	...	...	14.8–27.5	...	24.9–34.8	34.9–45.0	41.9–55.0	48.0–58.6	50.1–72.1	...
Mean	...	...	21.2	...	29.9	38.3	47.7	54.5	58.4	...
SD	...	...	8.99	...	7.00	4.14	4.60	5.00	11.96	...
SE	...	...	6.36	...	4.94	1.69	1.88	1.58	6.91	...
Foot length (fixed)	(N=2)	...	...	(N=1)	(N=1)	(N=1)	(N=4)	(N=4)	(N=2)	(N=2)
Range	5.4–10.5	...	...	...	...	...	47.5–72.0	72–80	70–78	75–78
Mean	7.97	...	...	...	...	...	60.7	74.7	74	76.5
SD	3.61	...	...	...	...	...	10.44	3.61	5.66	2.24
SE	2.55	...	...	...	...	...	5.22	1.80	4.00	1.58
Foot length (fresh)	...	...	(N=2)	(N=1)	(N=2)	(N=6)	(N=6)	(N=6)	(N=3)	...
Range	...	...	21.9–36.0	...	42.0–47.8	50.2–59.7	57.4–75.1	60.0–86.9	75.2–82.9	...
Mean	...	...	29.1	...	44.9	53.6	66.0	73.0	78.5	...
SD	...	...	10.00	...	4.09	3.58	6.88	8.93	3.87	...
SE	...	...	7.07	...	2.90	1.46	2.81	3.59	2.24	...
Tail length (fresh)	...	...	(N=3)	(N=1)	(N=2)	(N=6)	(N=6)	(N=6)	(N=3)	...
Range	...	...	50–82	...	95–125.9	124–150	125–190	147–220	175–230	...
Mean	...	...	63.4	...	110.5	135.7	155.5	180.3	205	...
SD	...	...	16.65	...	21.84	10.89	23.18	27.43	27.84	...
SE	...	...	9.62	...	15.44	4.45	9.47	11.20	16.07	...

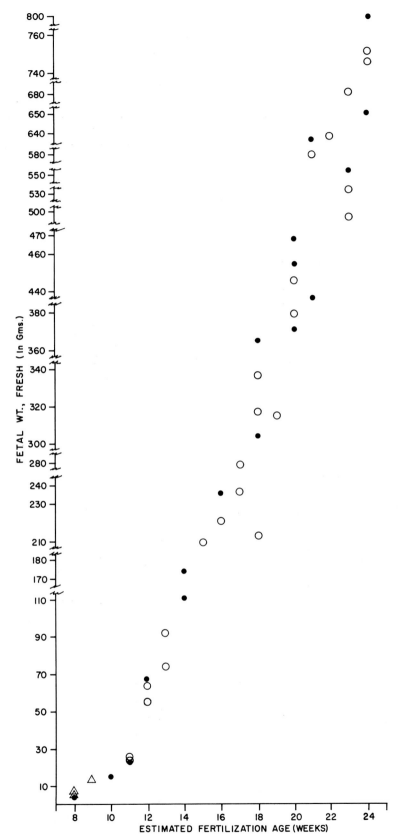

Fig. A.1. Body weights of males (●), females (○), and fetuses of unknown sex (△) with EFAs of 8 to 26 weeks.

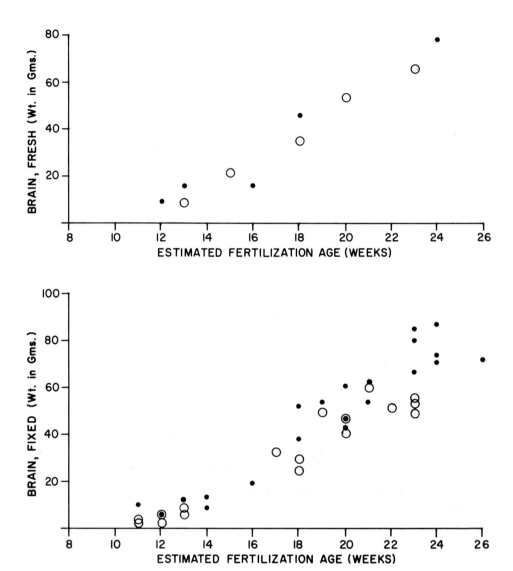

Fig. A.2. Weights of fresh and fixed brains from male (●) and female (○) fetuses with EFAs of 11 to 26 weeks.

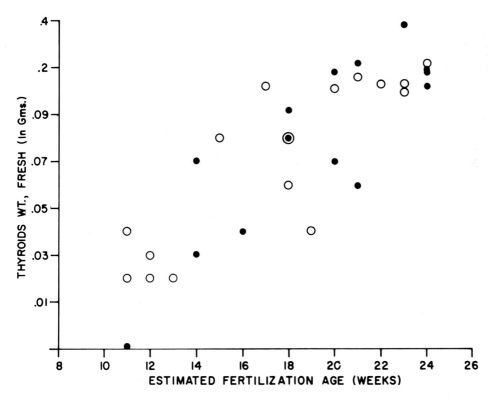

Fig. A.3. Weights of fresh thyroids from male (●) and female (○) fetuses with EFAs of 11 to 24 weeks.

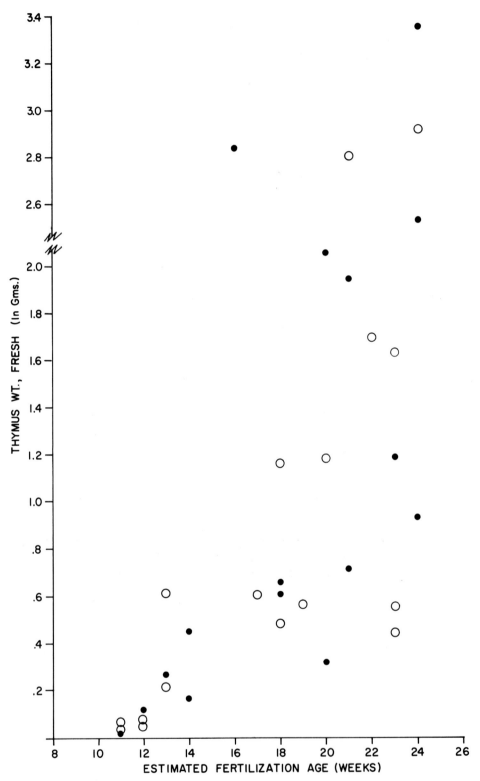

Fig. A.4. Weights of fresh thymuses from male (●) and female (○) fetuses with EFAs of 11 to 26 weeks.

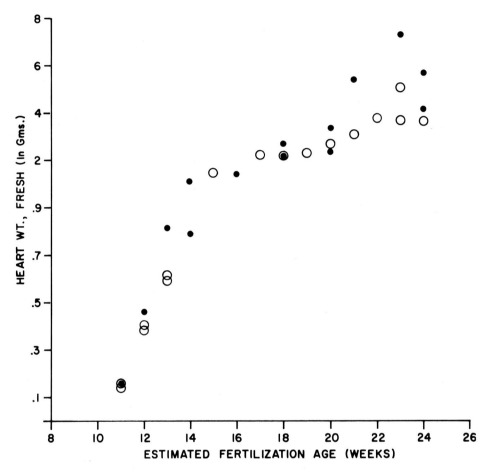

Fig. A.5. Weights of fresh hearts from male (●) and female (○) fetuses with EFAs of 11 to 26 weeks.

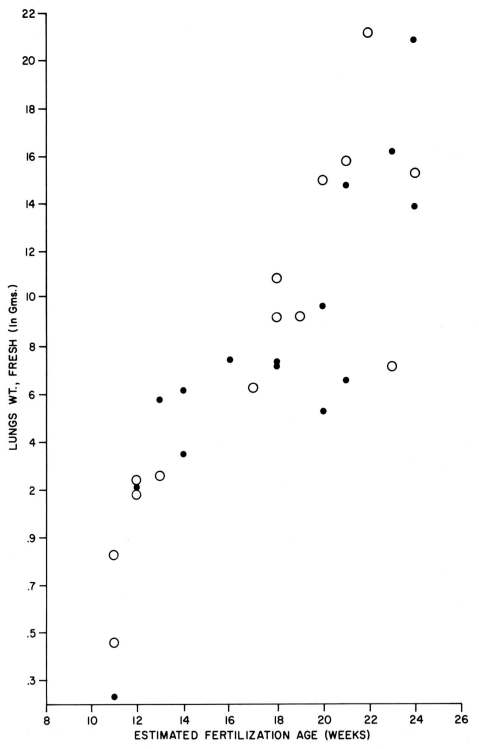

Fig. A.6. Weights of fresh lungs from male (●) and female (○) fetuses with EFAs of 11 to 26 weeks.

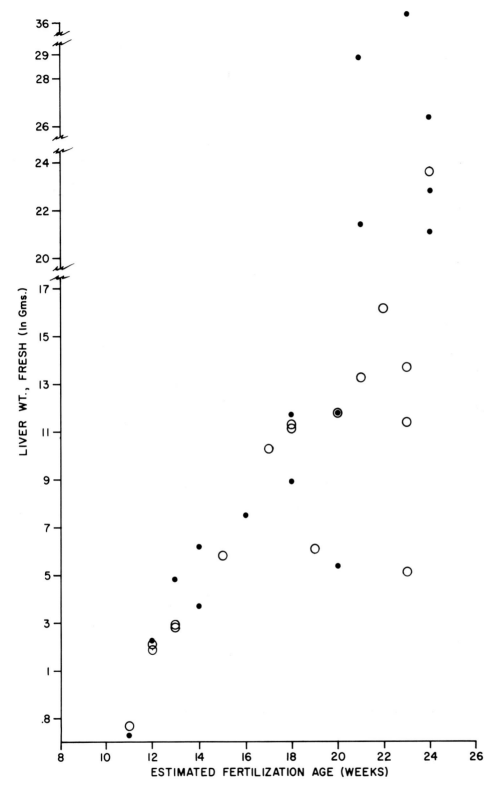

Fig. A.7. Weights of fresh livers from male (●) and female (○) fetuses with EFAs of 11 to 26 weeks.

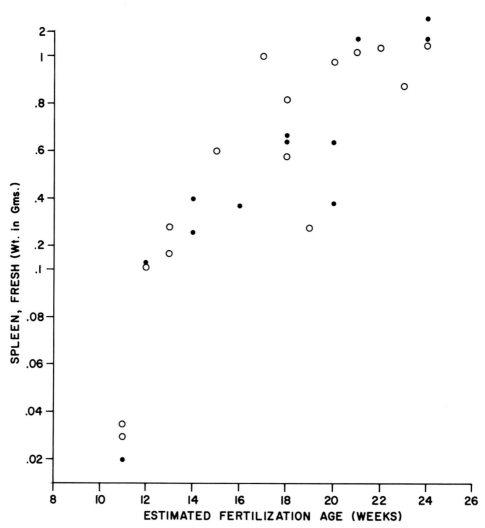

Fig. A.8. Weights of fresh and fixed spleens from male (●) and female (○) fetuses with EFAs of 11 to 26 weeks.

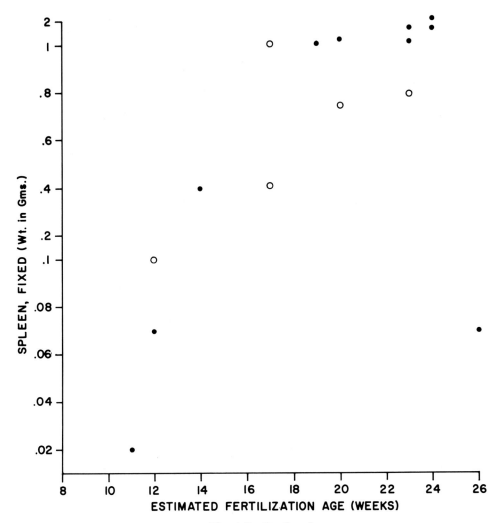

Fig. A.8—*Continued*

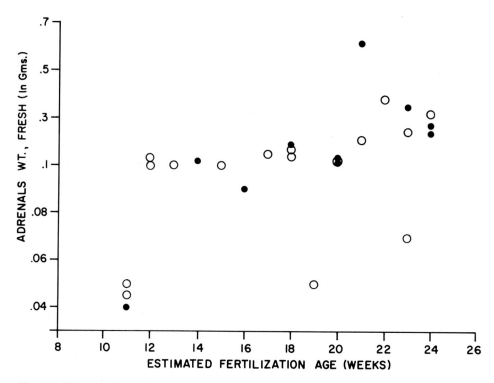

Fig. A.9. Weights of fresh adrenals from male (●) and female (○) fetuses with EFAs of 11 to 26 weeks.

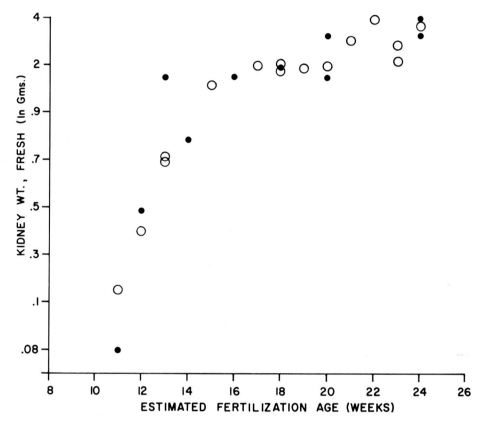

Fig. A.10. Weights of fresh kidneys from male (●) and female (○) fetuses with EFAs of 11 to 26 weeks.

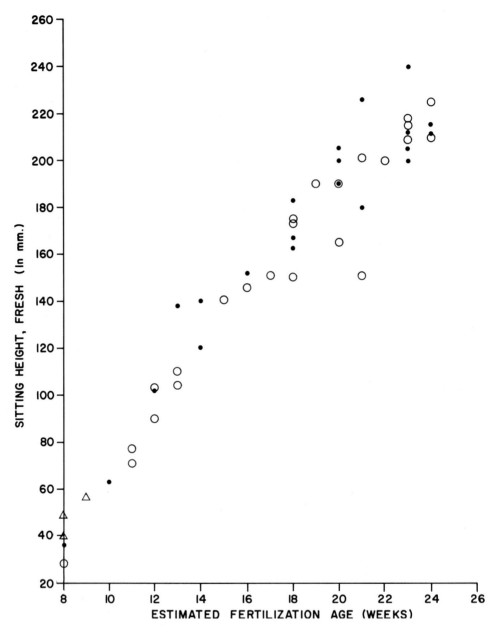

Fig. A.11. Sitting height (crown-rump) for fresh and fixed males (●), females (○), and fetuses of unknown sex (△) with EFAs of 8 to 26 weeks.

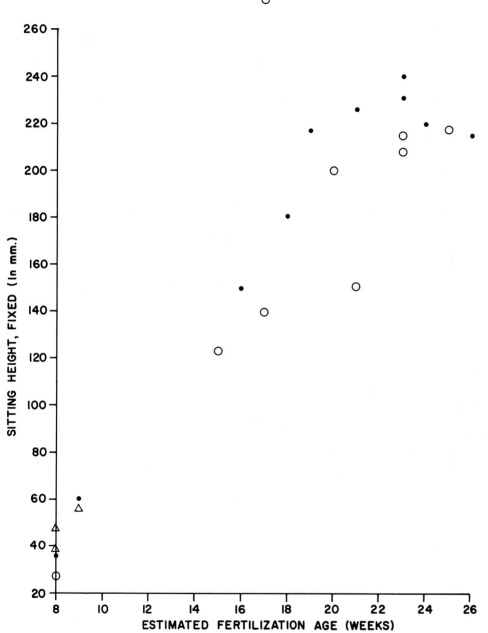

Fig. A.11—*Continued*

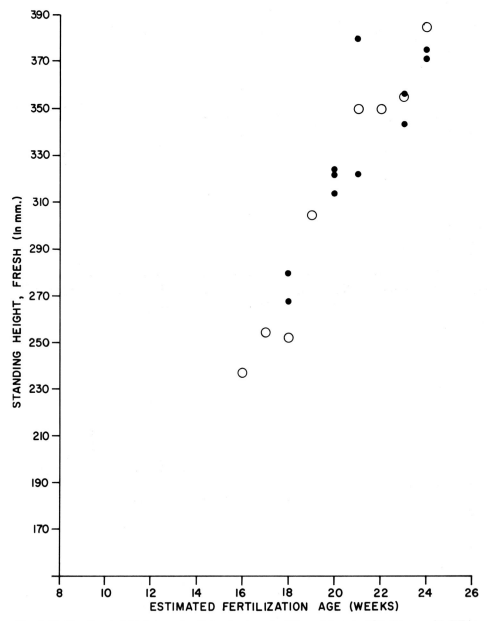

Fig. A.12. Standing height (crown-heel) for fresh male (●) and female (○) fetuses with EFAs of 12 to 24 weeks.

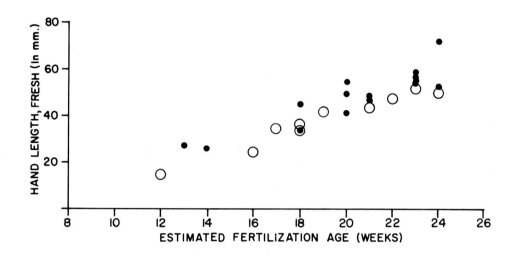

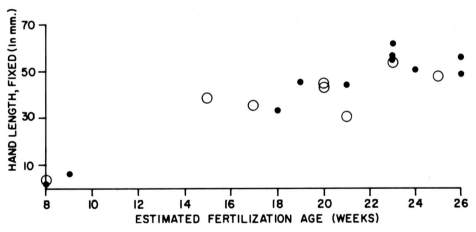

Fig. A.13. Hand length for fresh and fixed male (●) and female (○) fetuses with EFAs of 8 to 26 weeks.

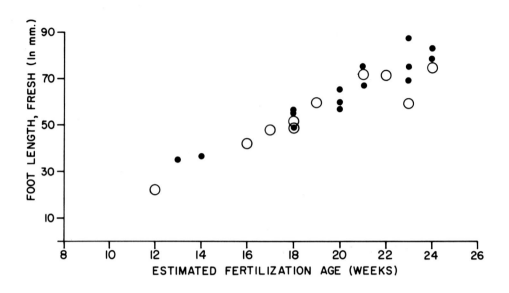

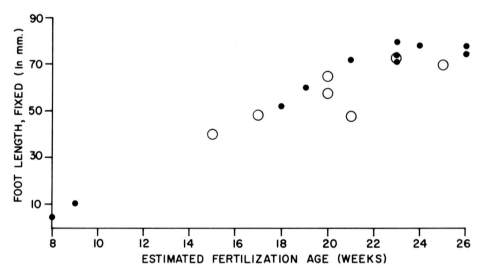

Fig. A.14. Foot length for fresh and fixed male (●) and female (○) fetuses with EFAs of 8 to 26 weeks.

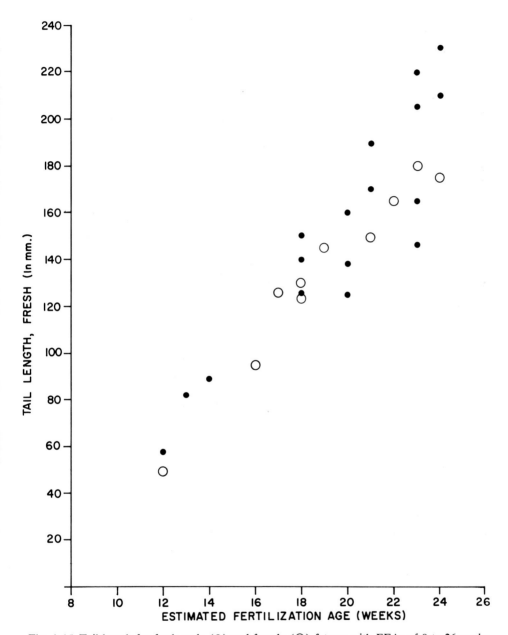

Fig. A.15. Tail length for fresh male (●) and female (○) fetuses with EFAs of 8 to 26 weeks.

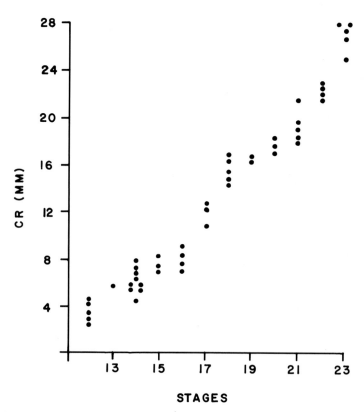

Fig. A.16. Crown-rump (CR) length of embryos belonging to Stages X–XXIII.

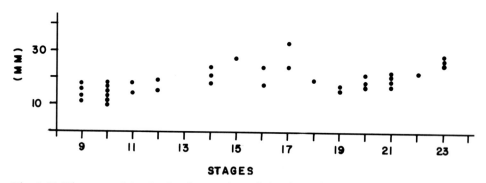

Fig. A.17. Diameters of the chorion from embryos belonging to Stages IX–XXIII. The chorionic diameters are plotted in millimeters on the abscissa. The chorionic diameters for Stages IV–VIII are given in table 4.2.

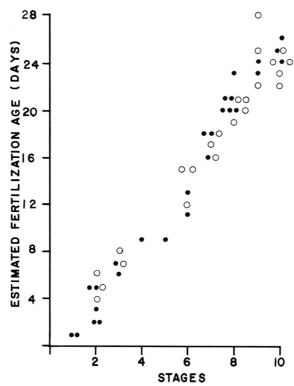

Fig. A.18. Estimated fertilization ages of embryos belonging to Stages I–X. Solid circles (●) represent embryos from single matings and open circles (○) represent embryos from continuous matings. The variation in age for embryos of Stages II and VIII exceeds by only 1 day the normal range for those stages. The ages of Stages I, III, VI, and VII embryos are within the normal range (chaps. 3 and 4).

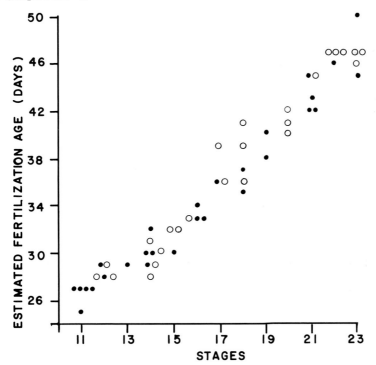

Fig. A.19. Estimated fertilization ages of embryos belonging to Stages XI–XXIII. Solid circles (●) represent embryos from single matings and open circles (○) represent those from continuous matings. Note that in Stages IX, X, XIV, XVII, XVIII, XXI, and XXIII where the age varies 3 or more days, the youngest or oldest embryo or both are the result of a continuous mating.

195

References

Kriewaldt, F. H., and A. G. Hendrickx 1968. Reproductive parameters of the baboon. *Lab. Anim. Care* 18:361–70.

Schultz, A. H. 1929. The technique of measuring the outer body of human fetuses and primates in general. *Contrib. Embryol., Carneg. Inst.* 20:213–58.

———— 1937. Fetal growth and development of the rhesus monkey. *Contrib. Embryol., Carneg. Inst.* 26:71–97.

Streeter, G. L. 1920. Weight, sitting height, head size, foot length and menstrual age of the human embryo. *Contrib. Embryol., Carneg. Inst.* 11:143–70.

Van Wagenen, G., and H. R. Catchpole 1965. Growth of the fetus and placenta of the monkey (*Macaca mulatta*). *Am. J. Phys. Anthrop.* 23:23–34.

Contributors

MARSHALL L. HOUSTON
  Department of Anatomy
  Division of Biological Growth and
    Development
  Southwest Foundation for Research
    and Education
  San Antonio

RAYMOND F. GASSER
  Department of Anatomy
  Louisiana State University School of
    Medicine
  New Orleans

DUANE C. KRAEMER
  Department of Reproductive
    Physiology
  Division of Clinical Sciences
  Southwest Foundation for Research
    and Education
  San Antonio

JOE A. BOLLERT
  Department of Anatomy
  Michigan State University
  East Lansing

Index

199